Management and Industrial Engineering

Series editor

J. Paulo Davim, Aveiro, Portugal

More information about this series at http://www.springer.com/series/11690

Carolina Machado · J. Paulo Davim
Editors

MBA

Theory and Application of Business and Management Principles

Springer

Editors
Carolina Machado
Department of Management
School of Economics and Management
University of Minho
Braga
Portugal

J. Paulo Davim
Department of Mechanical Engineering
University of Aveiro
Aveiro
Portugal

ISSN 2365-0532 ISSN 2365-0540 (electronic)
Management and Industrial Engineering
ISBN 978-3-319-28279-4 ISBN 978-3-319-28281-7 (eBook)
DOI 10.1007/978-3-319-28281-7

Library of Congress Control Number: 2015960406

Printed on acid-free paper

This Springer imprint is published by SpringerNature
The registered company is Springer International Publishing AG Switzerland

Preface

Thought to be one of the most prestigious and sought-after degrees all over the world, the Master of Business Administration, shortly known as MBA, is designed to develop the skills required in business and management careers. However, although directed to the business world, its value is not limited to it. Indeed, an MBA can also be very useful for all those that pursue a managerial career in the private industry, public sector, government, and technological and engineering area, among others. Being so transversal to these different areas, this is why in our days many academics and professionals desire to course and obtain an MBA. Conscious of this reality, this present book, entitled *MBA—Theory and Application of Business and Management Principles*, looks to contribute to some of the "core" curriculum of subjects usually present in an MBA program such as marketing, human resource, accounting, and finance. Based on the "core" subjects presented in this book, everyone interested will be able to obtain a relevant knowledge that can be applied as a whole to a variety of real-world business situations, or from a particular viewpoint that will allow them to follow their own personal or professional interests. Covering many different fields within business, the lecturers of this book, according to their interests and availability, will be able to obtain some of the most critical skills and knowledge subjacent to an MBA that will help them pursue a career in a variety of working fields.

Also providing a support to academics and researchers, the book focusing on the latest developments and thinking in what concerns the most recent research activity provides discussion and the exchange of information on principles, strategies, models, techniques, methodologies, and applications in the business area.

Following these concerns, this book, divided into three parts, covers the theory and application of business and management principles in six chapters. In *Part I—Speaking About Marketing*, the first chapter discusses "Marketing in Crises—Its Nature and Perspectives for Managers" and the second chapter contains information about "Understanding Digital Marketing—Basics and Actions." In *Part II—Speaking About Human Resource Management*, the third chapter covers "Human Resource Management: An Operational Perspective," and at the same time, the fourth chapter describes "Training and Development in Organizations: Start at the Beginning." Finally, in *Part III—Speaking About Accounting and Finance*, the fifth chapter focuses on "Accounting as an Information System," while in the sixth chapter, an "Introduction to Corporate Finance" is presented.

These theoretical and practical contributions will lead to an upper level of knowledge of these functional managerial and business subjects, and at the same time, it will contribute to the acquisition of new conceptual skills able to answer the challenges and changes set by the competitive business environment in which organizations are involved.

This book is designed to increase the knowledge and effectiveness of all those interested in the continual success of their careers in the different fields of the economy such as university research and activity (at the postgraduate level), business, manufacturing, education, health care, and other service and industrial sectors.

The editors acknowledge their gratitude to Springer for this opportunity and for their professional support. Finally, we would like to thank all chapter authors for their interest and availability to work on this project.

Braga, Portugal Carolina Machado
Aveiro, Portugal J. Paulo Davim

Contents

Contributors

Ana Alexandra Caria Department of Management, School of Economics and Management, University of Minho, Braga, Portugal

Ana Paula Vieira Gomes Ferreira Department of Management, School of Economics and Management, University of Minho, Braga, Portugal

Delfina Rosa Rocha Gomes Department of Management, School of Economics and Management, University of Minho, Braga, Portugal

Carolina Feliciana Machado Department of Management, School of Economics and Management, University of Minho, Braga, Portugal

Xabier Martínez-Rolán Faculty of Communication and Social Sciences, University of Vigo, Vigo, Spain

Leoni Eleni Oikonomikou Georg-August-Universität Göttingen, Göttingen, Germany

Lídia Cristina Alves Morais Oliveira Department of Management, School of Economics and Management, University of Minho, Braga, Portugal

Teresa Piñeiro-Otero Faculty of Communication Sciencies, University of A Coruña, A Coruña, Spain

Jochen Schellinger Business School, Bern University of Applied Sciences, Bern, Switzerland

Anabela Martins Silva Department of Management, School of Economics and Management, University of Minho, Braga, Portugal

Kim Oliver Tokarski Business School, Bern University of Applied Sciences, Bern, Switzerland

Part I
Speaking About Marketing

Marketing in Crises—Its Nature and Perspectives for Managers

Jochen Schellinger and Kim Oliver Tokarski

Abstract Corporate crises are recurring events in the evolution of almost any company. They are a regular part of a company's development, and it is part of any manager's job to deal with them. Against this background, this chapter explores a contingent approach to crisis-specific management of various marketing policies. The basic assumption is that situational, contingent properties of crises (should) determine the structure and behaviour of marketing management. Thus, managers should aim to achieve a specific "crisis-marketing fit" which involves the subfields of strategic marketing, and a set of typical responses for product, sales, pricing and communication policy.

1 Introduction

Since the turn of the millennium, more and more authors have been drawing attention to a dramatic rise in the economic dynamics, complexity, and intensity of competition for companies in the consumer and investment goods markets. Globalization, the advent of modern information and communication technology, and the changes in the values and expectations of consumers are often mentioned as the main forces behind this evolution. While the trend towards an increasingly complex and challenging commercial world has long been a general fact of life, the impact of the global financial and economic crisis in the wake of the collapse of Lehman Brothers in 2008 has only intensified these difficulties for companies and led to a spate of crises that often threatened the survival of the affected companies. Typically, these are highly dynamic, highly complex, and highly relevant for a

J. Schellinger
Business School, Bern University of Applied Sciences,
Brückenstr. 73, CH 3005 Bern, Switzerland

K.O. Tokarski (✉)
Business School, Bern University of Applied Sciences,
Brückenstr. 73, CH 3005 Bern, Switzerland
e-mail: kim.tokarski@bfh.ch

C. Machado and J.P. Davim (eds.), *MBA*,
Management and Industrial Engineering, DOI 10.1007/978-3-319-28281-7_1

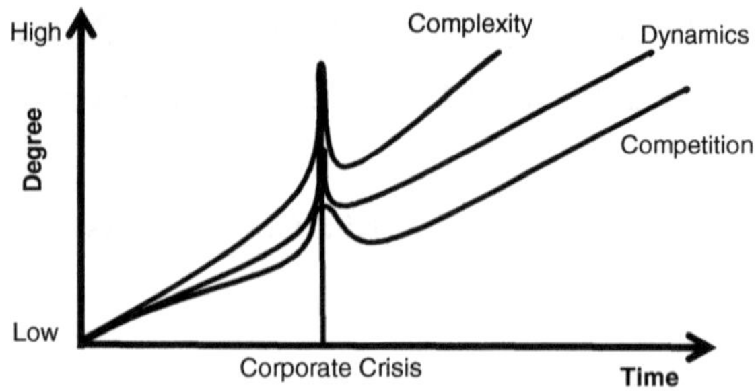

Fig. 1 Complexity, dynamics and competition in a crisis

company's strategic presence in its competitive environment and for the intensity of competition (Fig. 1). This is also reflected in the growing demand for guidance and advice on how to respond to this triple punch of dynamic, complex, and competitive challenges, as evidenced by the increasing number of publications on crises and crisis management. This new wave of research into corporate crises and their management has, as yet, not developed differentiated voices with pragmatic perspectives for procurement, production, marketing, human resources, or finances. This is where the following exploration of marketing and its specific opportunities for crisis management comes in. Publications in the field of marketing have mostly been interested in "crisis communication", which most authors consider particularly relevant for surviving and overcoming corporate crises. These questions of communication in crisis management need to be investigated, as due the questions of marketing strategy and product, pricing, or sales policy.

2 Crises as the Stage Setting for Marketing

In order to understand and describe the opportunities for marketing in the setting of a company crisis, it is essential to understand the terms, objects, and management-relevant aspects of a crisis in terms of the scenario it represents for the many perspectives of marketing policy. The key is the basic recognition of "crises" and their characteristic qualities as recurrent elements of considerable practical relevance in the cyclical development of any company. These recurrent crises call for specific management responses, which crisis-oriented marketing management needs to live up to (contingency approach).

2.1 The Objects of Company Crises

"***Crisis***" would seem to have become the dominant theme of the current times, with economies still reeling from the impact of the financial crisis, economic crisis, and Euro crisis or the many other crises of a more political nature (international terrorism, the Crimean crisis, the Syria crisis, the crisis of Greece, etc.). Crises in this sense refer to concrete threats that have become salient and urgent and that need to be mitigated or managed to protect the people affected from their potential impact. Crises are the result of prior risk scenarios; in this sense, they are risks made manifest. "Risks" are therefore the possible realization of unfavourable conditions which make certain hurt and damage likely. If such an unfavourable scenario has indeed become actual reality, the people affected need to contain its impact and stop the aftershocks. The general notion of the term "crisis" is therefore coloured by a risk-oriented macro-perspective that can just as well be applied to the micro-perspective of the affected institutions. One such micro-perspective is the treatment of the crisis phenomenon at commercial companies, that is, companies involved in a competitive market.

"***Company crises***" can similarly be defined as the occurrence of certain constellations of risks with a considerable, often existential threat and major strategic relevance for the company affected. Such crises, as already explained, are typically inherently dynamic, complex, and threatening to the company's competitive standing. This risk-focused notion of crisis is just one of many possible interpretations; it emphasizes the possibility of anticipating and influencing crises before they hit and their strategic (planning) relevance. In practice terms, it brings together the perspectives of risk and crisis management. German literature has increasingly become reliant on a property-oriented procedural definition of crises [1], which would consider company crises as unintentional and unplanned processes that are temporary, only partially susceptible to influence, and ambivalent in terms of their outcomes. Crises can endanger the continued survival of a company by impacting on its goals [2, 3]. They typically occur with little or no forewarning and create significant pressure in terms of time, stress and their impact on the ability to fulfil the expectations of relevant stakeholders [1]. In the four German-speaking countries of Europe, media reports suggest that every year witnesses about 220–260 such critical incidents in private enterprises or official institutions [4]. In 2014, more than 13,500 bankruptcy proceedings were begun in Switzerland [5], compared to 24,000 in Germany [6] and 5400 in Austria over the same period [7]. The collapse of companies can be considered the negative end point of a corporate crisis. This suggests that even the considerable presence of such crises in the media reflects only a minor share of the actual incidents. At the same time, this implies that company crises are an essential and existential object of everyday business operations.

2.2 Types and Causes of Crises

The most essential thing to do when trying to affect the course of a company's crisis is to consider and understand its cause and nature, which often affect each other. ***The causes of crisis*** can be many and diverse: researchers have identified a vast number of relevant constellations from their quantitative and qualitative studies. The first avenue of research is aimed at typical crisis-prone or crisis-favouring properties of companies in crisis. Companies in the service sector, small businesses and smaller limited companies, companies established for fewer than four years and older companies of 10 years or more are particularly prone to crisis [1]. The basic distinction between exogenous and endogenous occasions of crisis goes back to the original research into the qualitative causes of crisis [8]. Hauschildt [9] argues against this distinction, as the responsibility for a crisis and for managing the crisis always lies within the organization in question or, in other words, with "mismanagement", even if exogenous factors come into play: "*crises are caused by the failure of management to adjust adequately to external circumstances, not by these circumstances as such*". Even if the distinction between internal and external causes holds, researchers commonly point to the dominance of internal forces or the typical finding of a combination of internal and external factors [10]. Assuming in this sense that the causes of crises are essentially always to be found in management, which relates very well with a strategic understanding of the phenomenon, we can distinguish between different elements of crises and their relevant causes by distinguishing between different areas of practice failures and causal forces [9]:

- The entrepreneur as a person, e.g. inexperience, poor leadership, illness.
- Institution/Organizational state, e.g. strategic mistakes, poor choice of legal form, problems with associated partners, unhelpful organizational structure, poor information management, poor HR management.
- Financial, e.g. credit lines cut, receivables lost.
- Operational causes: poor decisions about investments, productions, logistics, procurement or sales.

Considering the ***causes of crisis in marketing***, Hauschildt explicitly refers only to operational causes concerning the product range (outmoded technology, too high/too poor quality, no thought-through portfolio, internal competition), pricing policy (too expensive/too cheap, not dynamic, not sustainable) and sales channels (too strictly/loosely defined, too abrupt changes). Problems in communication policy (e.g. excessive advertising spending and failure to reach the right audience) are not addressed by him. Such aspects relating to marketing strategy are related to the institutional/organizational failure cluster (too fast external growth, too strong reliance on a too powerful partner). The list of causes mentioned here is therefore clearly not complete. In an earlier study of the relative frequency of the different failure types, Hauschildt [11] names sales problems as the most common crisis-causing type of failure. More recent empirical studies also point to the great importance of sales in this respect. Hauschildt et al. [12], for instance, attribute more than 20 % of all company crises to the aspects of sales or the market state. A similarly dominant position is held only by leadership failure.

Just as researchers have been distinguishing between different categories of causes for crises, different concepts for distinguishing different ***types of crises*** have also been put forward. The criteria for making such a distinction include, in particular, threats to the company's goals (value, material or social goals), the relevant causes (internal, external or combined causes), the incidence of the crisis in the company lifecycle (growth, stagnation or contraction crises), the dynamic state of the crisis (threatening, latent, acute, but manageable, acute and uncontrollable), or the time available to prevent or mitigate the crisis (strategic crisis, performance crisis, liquidity crisis, insolvency) [10]. These types are not without their overlaps and similarities and can clearly be combined with each other. Depending on the type of crisis, different forms for preventing or mastering it are thinkable: Müller's [13] basic distinction between strategy, performance and liquidity crisis in particular has proven its practical worth as an empirically sound patterns for structuring the crisis experience from a managerial point of view. Strategic crises affect the capabilities of the organization in question to a massive degree, but they allow substantial leeway to prevent or mitigate the escalation of the crisis. Such strategic crises can evolve into performance crises over time and make themselves felt in many important performance indicators (e.g. profits, revenue). Liquidity crises represent the escalation of financial problems in a crisis scenario, affecting the organization's ability to service its liabilities and adding new debts, which can end in the final crisis phase of insolvency. A similar phase model of Töpfer [14, 15] excludes this final stage and contrasts the tripartite concept of "evolutionary crisis" (strategy, performance and liquidity crisis) with "sudden crisis" which erupt without forewarning and have an immediate impact on performance (e.g. insolvency of a main client, losing an important revenue stream). Figure 2 integrates these endogenous or exogenous shocks into the three-phase concept. In this sense, spontaneous crises are not seen as the opposite of the three-phase concept, but as shocks that can affect all active levels and can ramp up the urgency of the crisis massively. Spontaneous crises can lead to strategic, performance or liquidity crisis at any given point in time. The three-phase model is therefore not (just) an ideal sequence, but a representation of the different spheres for managerial action that remain relevant throughout the entire crisis period (and have no clear-cut borders between them). Each has its own urgency and calls for a specific response.

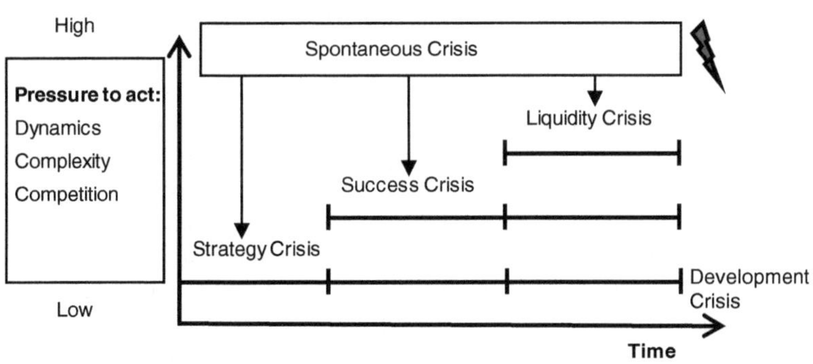

Fig. 2 Types of crisis and pressure to act

2.3 Crisis Management and Business Development

"Crisis management" has become the accepted term for the management and development of the corporate sphere before, during, and after a crisis. Its conceptual focus, however, lies clearly on overcoming crises that are active and have been recognized as such. This bias can also be seen in the general thrust of scientific involvement with this aspect of management.

Crisis prevention management has the basic purpose of either preventing crisis from occurring in the first place or, if that be impossible, to prepare the company as effectively as possible for responding to crises when they do occur [15, 16]. A more expansive notion of crisis prevention would include post-crisis activities after a crisis has been successfully mastered, which includes analysing and assessing the effectiveness of the crisis management processes. "Learning from the crisis" is the key here and helps create awareness for other potential crisis factors. It can professionalize decision-making processes in active crises, e.g. by teaching the company that a liquidity crisis needs to involve more stakeholders in communication than just the capital investors. Prevention correlates strongly with the strategic vision of crisis management. Crisis prevention management is an essential component of ***strategic crisis management***, which includes the management of strategy crisis and the strategic aspects of the operational response to performance or liquidity crisis. In terms of strategic management, crisis prevention management primarily concerns strategic monitoring [17] (strategy controlling, controlling the strategic premises, planned progress and potential). Considering the risk-oriented definition of crises and the great relevance of risks and opportunities in strategic decisions, risk management can also be included in crisis prevention as part of strategic crisis management. ***Operational crisis management*** begins as soon as a crisis strikes and covers all responses to prevent the collapse of the business. Both strategic and operational crisis management require ***functional specification***; that is, it needs specific monitoring and response plans for procurement, production, marketing, HR and finance. For marketing, this means that strategic and operational crisis management typically calls for action on the marketing strategy and product, price, sales and communication policy. Figure 3 presents the correlations and shared aspects of the different dimensions of crisis management.

Crisis management has also evolved into a distinct management function since crises have stopped to be considered as isolated, singular challenges for management and have become recognized as specific constellations of companies and their environments that can occur again and again in a ***company's evolution*** and are indeed important for the company's very ability to evolve. Crisis management, seen in this vain, represents company management approached from the vantage point of specific development phases. The need for and empirical presence of such a phase-dependent management approach are supported by several representatives of lifecycle-oriented management concepts [18] and is, as such, one pillar of the St. Gallen management concept [19–21]. The same can be said of the great relevance of crisis constellations for the qualitative and quantitative development of

Fig. 3 Perspectives of crisis management

companies and their markets. Greiner's model of growth is one of the most popular concepts in this respect, linking crises and the stages in a company's development by causal mechanisms [22]. Considering the special case of marketing, one can point to the five-phase model proposed by Becker [23]. Referring to the life cycles of companies, he sees the need for a new look at the basis questions of how marketing is done and distinguishes between the life phases of foundation, growth, saturation, and contraction. Each phase has its typical practices and leadership qualities (phase-oriented management types). Saturation and contraction represent more or less pronounced crisis phases, which often lead to a complete rethinking of marketing concepts in the pursuit of a turnaround. In them, the call goes up for "restructurers" or "pioneers" to initiate and steer the necessary change processes. The market and customer perspectives of the changes plays a particularly prominent role in this, "because only companies that align all of their activities with their markets and customers (market- and customer-driven companies) can survive in the long run" [23]. The all too frequent simplistic focus on reducing costs is not sufficient in this respect. Profit margins need to be targeted just as well by selling the right products to the right customers at the right time and for the right price. A fitting, crisis-oriented marketing mix is needed [23]. Becker's marketing concepts pay attention to the marketing-specific aspects of development-driven crisis management only to a rudimentary degree and without much more insight in terms of marketing policy. However, his model remains a very effective means for understanding the development dimension of crisis management. It matches the proposed notion of crises as recurrent normal states in which marketing plays a particularly important role: "… crises are becoming an inseparable part of modern business" [24]. The model also offers a first frame of reference for marketing practice that is aware of the crisis context and reminds us of the need for functional differentiation in crisis management. The main elements of Becker's proposals are illustrated in terms of their relevance for marketing in Fig. 4.

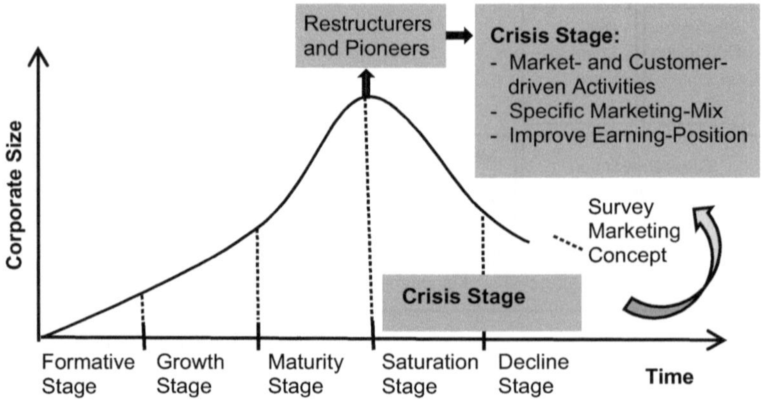

Fig. 4 Corporate development and the marketing aspects of crisis management according to Becker [23]

2.4 *The Contingency Approach: Marketing in the Crisis Context*

The need for a discussion of marketing in the context of company crises can be explained with the ***situational approach*** of organizational theory and the closely related "fit" philosophy of strategic management [17]. The basic idea of the situational approach suggests that the structures and systems that determine the efficiency of a company are formed by specific situational conditions [25–27]. From a managerial point of view, this means that a given endogenous or exogenous context would typically have an "optimum fit" combination of structures and systems, which affects behaviour and practice in such a way as to lead to effectiveness, efficiency and, by implication, profitability. The company and its environment are brought into the right fit by "contextually appropriate" (strategic) management (organization—environment fit). Taking company crises as temporary, but highly relevant and impactful contexts, their specific properties allow us to define specific requirements for management or, specifically, marketing from a contingency-driven, situational point of view. A number of recent marketing publications have taken up this train of thought, at least implicitly, when looking at the increasing frequency and relevance of critical constellations [24, 28–31], but similar ideas were also already included in older contributions, such as Shama's insights into the links between economic crises and marketing management [32]. On the strategic level, the contingency approach mirrors the idea of strategic management as a "dynamically flexible concept" [33]. In this sense, Ansoff's original proposition of strategic issue management and his concept of weak signals [34] can also be considered a fit-oriented crisis prevention or preparation technique. According to Lombriser and Ablanalp [33], crisis management itself is also part of such a dynamically flexible understanding of strategic management which always comes into play when pre-emptive measures have failed.

Figure 5 represents the basic mechanisms of a ***contingency-oriented marketing approach*** as a pragmatic application of the situational approach.

In terms of their relevance for practice, crises present two sets of characteristics: first, there are their general traits, such as their complexity, dynamic nature and competitive intensity; second, distinguishing different types of crisis is important to choose the right means for the context in order to master them. Crises are often affecting the conditions in the market and can influence the perceptions and habits of consumers or lead competitors to change their practices. The typical outcome is a drop in demand [35], which represents an important foothold for marketing in its response to the crisis on the sales side. Changing conditions in the market are often the results of the original causes of the crisis (e.g. the downturn in demand in many industries in the wake of the financial crisis), even though they tend to stand at the beginning of the causal chain for a company's specific crisis experience (e.g. a product crisis caused by reports of health hazards). As the relevance of such market-side effects of a crisis increases, the importance of marketing for overcoming the crisis also increases. Companies confronted with such crises need to

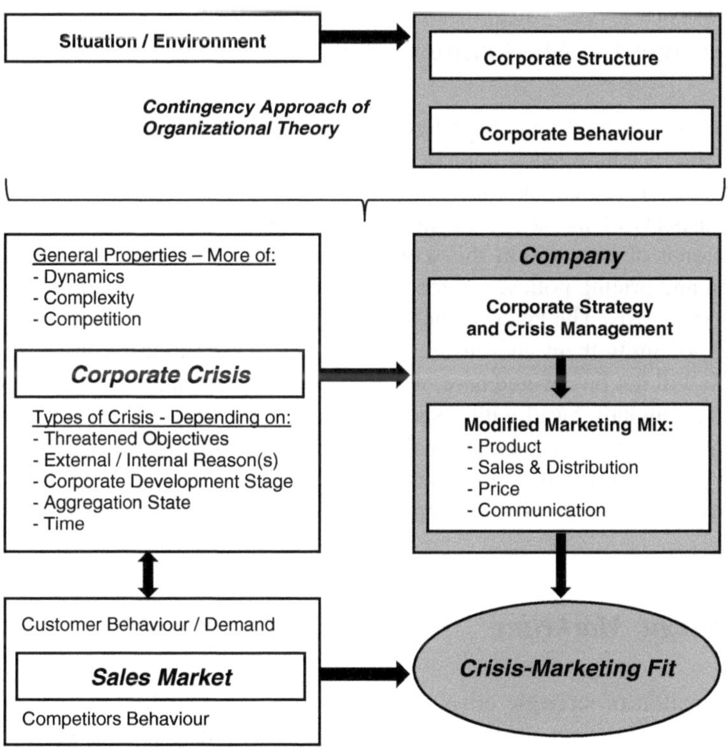

Fig. 5 Contingency approach of marketing in the context of corporate crisis

reconsider the impact on the corporate strategy and install crisis management that leads to changes in the marketing strategy and product, pricing, sales or communication policy. A marketing mix modified and focused for the crisis can contribute to achieving the right situational fit between the company and its new environment. The model can also be used to systematize possible options for preventing or preparing for crises (e.g. modelling crisis scenarios).

The following sections will investigate the various practical dimensions of the marketing mix in terms of adequate crisis responses. Following the contingency approach for crisis marketing as proposed here, the focus will lie primarily on understanding the types of crisis with strong direct or indirect ***effects on sales market***. This type of crisis will, in essence, always concern massive and disruptive drops in demand, which typically goes hand in hand with a change in competitors' practices before or during the crisis. Similar marketing responses would also be possible in other types of crisis whenever a sudden impact on revenues occurs as a result of certain critical constellations on the side of resources or capacities (e.g. critically high damages to be paid after infringing a patent).

3 The Scope for Marketing in Times of Crisis

As the crisis-specific frame of reference for marketing has been defined and translated into a contingency approach for marketing in times of crises, the next step, the second degree of concreteness, would be the definition of the most important aspects of the scope for marketing in practice. Before considering the four traditional objects within the scope for marketing, i.e. product, sales, communication and pricing policy, we need to consider the basic thrust and planning of marketing in a crisis. The latter defines the strategic frame of reference for the areas of practice, delimits them and ensures their coherent definition. A third and final step, which will not be covered here, would be the concrete application of all of this at a specific company faced with a specific crisis (case study). There are empirical studies that have investigated such real phenomena (also) in terms of marketing in crisis and proposed specific responses [24, 30, 36–40]. Their insights have been used to arrive at the following concepts on this second level of concreteness.

3.1 Strategic Marketing

The investigation of strategic crisis management has revealed three distinct components that are relevant on the strategic level: preventing crises, which covers strategic controlling and risk management; responding to strategic crises in terms of the threat to revenue streams; and considering the strategic aspects of the operational response to the crisis. Strategic crisis management therefore means more than overcoming or preventing evolutionary crises; it is also of relevance in unexpected, emergent crises. These distinctions also point to the most important pillars for strategic marketing in the crisis context. The three components now need to be filled with life in terms of marketing. Following the concept of market-oriented business management [38, 41], we can assume that the ***marketing strategy*** plays a vital role for the overall strategy of the company. In many cases, the marketing strategy is practically identical with the corporate strategy [42–44]. This applies in particular to cases where a more expansive understanding of marketing is used, which also includes stakeholder management and internal processes and resources in the design of the marketing strategy, which is often the case at small-to-medium-sized enterprises in particular. The important factor here is the underlying planning process. ***Strategic marketing planning***, its operational implementation in the marketing mix and its budget need to be flexible in a time of crisis. By implication, this also goes for the marketing strategy as a whole or in its various parts. Changes to strategic decisions are, however, not necessarily required in every crisis, as the changes can often be limited to temporary operational aspects in the various policy areas, e.g. to communications. One example of this would be a turnover crisis that leads to serious liquidity issues and might be the result of external sabotage, but has no relevance whatsoever to the actual marketing strategy or the business model it is based on. Apaydin and Geçti [30] emphasize the relevance of focusing on core operations and improving marketing efficiency as the key jobs for strategic marketing planning. Market monitoring is particularly important here, as is the (re-) planning and definition of market opportunities and the marketing mix, the stabilization or even growth of the company's market share, and the reviewing of the return on marketing investment.

When strategic marketing is concerned, ***three areas*** are relevant for ***investigation and practice***: the business model, the customer focus and the marketing strategy as the sum of the parts of the marketing mix (product, pricing, sales and communication strategy). Questions of marketing efficiency also need to be asked with a view to the costs involved.

The ***business model*** represents the core of the company and marketing strategies. It defines the most important pillars for the marketing strategy and expresses the company's potential for commercial success in a nutshell. Built for the long term, it should not, however, be static when one remembers the dynamic evolution of the business. In a crisis context, the dynamic changes under way reflect the innovations that are required [24, 31, 38]. Massive disruptions in the market can even call for a complete rethinking of the business model, e.g. when unexpected technological innovations make an old product obsolete overnight. Another such situation would be the forced entry of powerful and innovative competitors into one's home-turf market, which companies such as Google or Apple are wont to do. A strategic crisis also demands a replacement or modification of the business model. Depending on the main causes (endogenous/exogenous, resources/markets, customers'/competitors' practice), the focus of the adjustments changes. Nonetheless, there are few strategic crises that will have no impact on sales markets or customers. A procurement crisis can be caused by a competitor taking over the main supplier for essential capital equipment, which forces a company to change its product range or eliminate a product line or which causes massive problems with products and maintenance services down the line. The situation is different in more operational ad hoc crises that have little to do with the business model as such (e.g. disasters). Here, the focus will lie on adjusting the marketing mix on the ground. In a company crisis, all components need to be investigated to understand whether and how they concern the business model. The business model also offers a helpful means for analysing the monitoring and risks management measures in crisis prevention. The focus would here lie on market-focused considerations about possible changes in customers' preferences or competitors' decisions.

The ***customer focus*** represents an important point of reference for strategic marketing [43] at most companies. It means the coherent alignment of all products and services and the underlying processes with the needs and expectations of the customer. Neglecting that perspective is often the immediate or a contributing factor for a strategic crisis, which can then deteriorate into a performance or liquidity crisis. Two customer-related aspects are particularly relevant here: customer loyalty and profitability. Depending on the changing preferences of customers in a crisis or the actions of competitors, the basic activities that the company uses to ***retain established customers*** [28] need to be analysed, because the slow loss of formerly loyal customers will have substantial repercussions that go beyond the immediate loss of revenue. Every lost customers means more pressure to acquire new business, which is particularly problematic in a crisis and considerably more expensive than maintaining good relations with established partners. Apart from customer satisfaction, this calls for attention to other areas, such as changes to contractual terms to gear them more towards long-term relations or mutual dependence. However, customer retention is not enough for a customer-focused response to a crisis. Retention has to be selective. The ***profitability of the customer*** has particular relevance in a crisis. Customer retention has to try to maintain or promote only the profitable or commercially promising customer relationships. Unprofitable relationships should be reviewed in terms of their potential promise and, if there is no such potential, cut-off. This naturally also applies to new clients acquired in a crisis. Expanding market share by simply acquiring new, but unprofitable customers will only reinforce the crisis. Neglecting the question of profitability is a common reason for why companies fall prey to crises and fail to survive them. The emphasis on customer focus as an important strategic reference point has many implications for the various dimensions of the crisis marketing mix, which will still be investigated. The basic principles that underlie the question of customer focus also need to be considered when trying to prevent a crisis.

Marketing strategies are functional expressions of the corporate strategy that go beyond defining the sales market, although this stands at their centre. They represent pragmatic guidelines for implementing the company's and marketing targets. The marketing strategies are typically developed on the level of the strategic business units (one or more related product/market combinations) [41]. The frequent identity with the corporate strategy has already been mentioned. Lifecycle-oriented standard strategy models are intended to come up with phase-specific recommendations for shrinking or stagnating markets and focus on added value, customer loyalty measures, rationalization and process streamlining. The option of withdrawing or exiting from a market is always included by implication if remaining in the market is not possible or not a viable option [41]. Current literature also distinguishes between a range of ***market strategy types*** that relate to type-specific questions [42] and represent the relevant strategic decisions. They become the strategic bundles that make up the holistic marketing strategy. Meffert, for instance, distinguishes between the following types [41]:

- Market selection strategies (market field strategy = product-market strategies; market area strategy = regional definition; market segmentation strategy = differentiation in market approaches);
- Market actor strategies (customer-oriented = developing the market in terms of the customer; retailer-oriented = retail strategy; competition-oriented = competitive strategy; stakeholder-oriented = indirect market management).

In essence, when a crisis occasions a company to rethink its strategic decisions, the company needs to check and, possibly, revise their quality in practice and their general validity. This can be done by means of a business model analysis and by processing the results of a customer focus analysis, paying due attention to the interrelations and correlations in strategic company planning and the company strategy. The strategic framework chosen for the marketing mix instruments (product, sales, communication and pricing strategies) needs to be adjusted where necessary and implemented in operational practice [30, 37]. This is affected by the crisis' typical impact on available resources (e.g. high costs) and the mentioned characteristics of crises. As in business model and customer focus analyses, the parameters for strategic decisions also need to be referred to when monitoring and scanning for potential strategic discontinuities and risks. In addition to these three main objects, the actual marketing activities of the company need to be checked and optimized in terms of their efficiency. Sophisticated cost-oriented ***efficiency measures in marketing*** should definitely be preferred to blanket cuts in marketing budgets which are often the first option that companies go to, aside from unthinking price reductions. The trick is not to reduce the activities in general, but to make more efficient use of the available means in the crisis.

Figure 6 sums up the main aspects of strategic marketing in a crisis.

3.2 Product Policy

Product policy concerns the programmes and measures that are normally determined by product strategic considerations. It concerns practical decisions about individual products, product lines or entire product ranges.

For product policy, the following parameters of product strategy [41–43] are particularly relevant and should be included in the often short-notice reviews and changes when a crisis has struck:

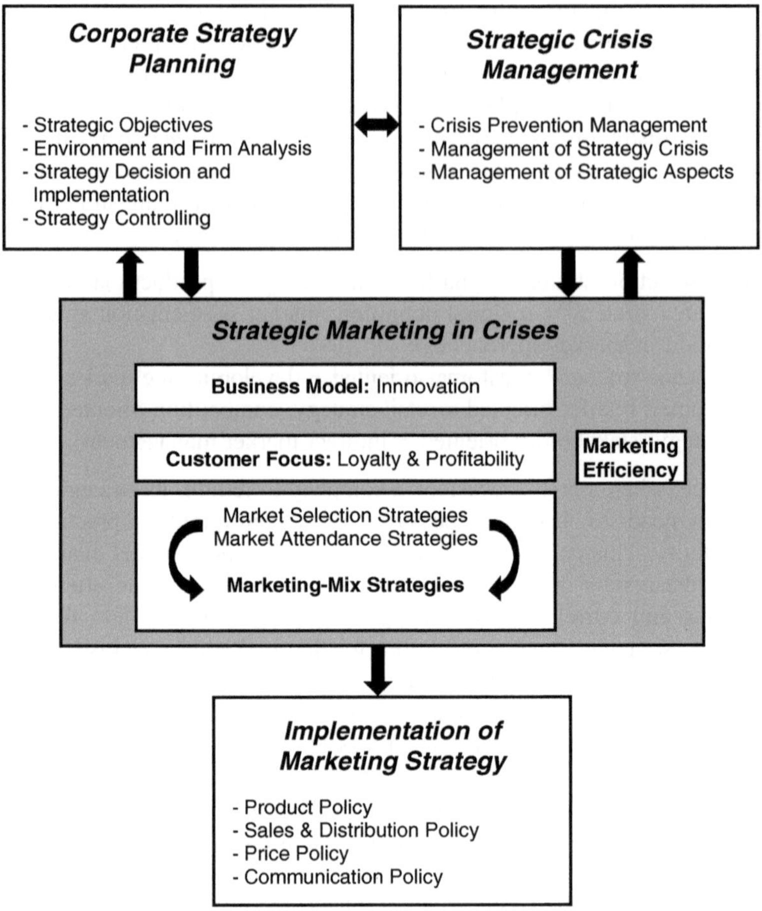

Fig. 6 Strategic marketing in a crisis

- Product quality, features and design (functionality and value for the customer).
- Brand and packaging.
- Product-related services and warranties.
- Product lines and product portfolio.

Product policy covers many dimensions: product innovation and development, variation and diversification, or product elimination. Brand management, that is, the focused management of the brand's value, is another related dimension [42, 43] that determines the nature, reach and structure of the brand, the brand presence (in particular, the brand name), intellectual property or other aspects of brand development [41–43].

A drop in demand and intensification of competition as the result of a crisis, leading to changing habits among customers and competitors, calls for a number of

product policy responses. In current literature, product policy and other aspects of marketing policy are discussed as well-established and empirically proven factors for commercial success, with specific options for responding to the typical lower demand and scarcer resources in a crisis. This can be seen in the publications of Marasović et al. [28] and Simon [45].

The first perspective for product policy lies in ***tightening and standardizing the product range*** to become less complex [30, 39, 46] and allow the company to concentrate on products that are sufficiently or even highly profitable. In turn, products that are not commercially rewarding enough should be eliminated if this is possible in view of the repercussions for the product portfolio. This streamlining and standardization—or indeed modularization—are often neglected options in crises, although they have great promise [46]: "*our experience suggests that changes to the product range (be it to sell more or to save product costs) are the greatest lever for a turnaround. It is surprising to see how rarely this lever is used in restructuring*". Adding a ***new product*** should only be an option when it represents an essential expansion of the core range. Integrating vertically "neighbouring" products in the range offers the opportunity to evolve from a product manufacturer to a provider of systems [45]. The pressure on resources in crises leads many companies to cut back their own research and development activities in terms of product innovation [28], which is a clear case of saving at the wrong end. There is empirical evidence to suggest that ***increasing research and development spending*** is much more promising for crisis management, irrespective of the type of company in question [24, 47]. There are also definite signs that involving the customer in the innovation process is a positive idea, not least in a time of crisis. Intensive and positive customer relations are a key resource for added value, again also in a crisis [35, 48, 49]. Another common recommendations for improving the appeal of one's products is to ***offer more services*** around the product affected by the crisis. Additional, expanded or more flexible services in demand by customers can have a significant impact on commercial performance. These services often have much higher profit margins than the original products in question, are less cyclical in nature and can thus cushion the impact of the crisis. They can also help retain customers through the crisis and are a good example of the customer focus principle of crisis management. They make the company's competitive advantages more durable and make its offerings less easily replaceable. Another option in this respect is the offering of unusual warranties and guarantees to reduce the risks for the customer (e.g. guaranteed returns on real estate investments) or of extended trial periods [28, 30, 37, 45]. As consumers are afraid and more careful with their money in a crisis, as was seen in the global financial crisis and its aftermath, companies also need to be aware of the changing habits of consumers in terms of different ***product categories***. In the years after 2008, the markets revealed interesting differences in consumers' habits: Essential products (such as drugs) were not affected at all. Products that are deemed necessary and whose purchase cannot feasibly be delayed (e.g. food or hygiene articles) were only affected to a minor degree. The market for so-called postponables, that is, for products that could just as easily be purchased at a later date (e.g. cars) or luxury goods (e.g. designer clothes) suffered from a downturn in demand in the high double digits, bringing many companies to the brink of collapse. There were other market segments that indeed benefited from the crisis (such as used cars) [28]. A detailed sense for the different preferences of customers can be essential for product policy in a crisis. One example of a response to the crisis of 2008 was the accelerated allocation of company cars to the employees of the luxury car maker Daimler AG, which fed into the used car market for the company's models. All of the measures named up to now are aimed at increasing gross turnover or reduce costs. In some instances, however, it can help to limit the total sales volume, as a drop in the volume often has a less severe impact than a drop in prices. ***Sales volume management*** can keep prices stable and shield against competition. However, it is only feasible if the competition behaves likewise, which suggests that the market needs to be closely monitored and responded to [45].

Brand management faces the challenge of realigning existing brands with new or changing preferences in the market. In a general downturn in demand as a result of an economic crisis, the emphasis should lie on the value for the customer. Consumers also tend to give their preference to local or domestic products in such instances, which could be an opportunity to emphasize the (local) origins of the product. Another tendency among consumers is to replace premium brands with less popular or store brands [28], which allows companies to upscale their brands and poach market share from formerly dominant premium competitors. A strong brand is an important source of distinction in a crisis, with its emotional charge contributing substantially to maintaining or even improving turnover. It also offers opportunities for entering into new market segments by diversifying the brand. Identity-driven brand management tries to establish a strong brand and therefore has great potential for preventing or preparing for eventual crises [37, 50–52]. Changes in the product's or brand's perception can, however, also be a cause for a reputation crisis, which places brand management at the frontlines of the crisis response [53]. Brand management in a crisis is, in this sense, closely linked with crisis communication, and both policy areas need to be fully aligned.

Recommendations for the ***maturity or degeneration phases*** of products' life cycles [43] mostly concern market development measures in the sense of repositioning the brand, developing new target groups or increasing the usage. For product development, the focus here shifts to quality improvement, added functionality, new variants and models, or new designs. Figure 7 presents the main product policy responses for implementing the product or portfolio strategy in a time of crisis.

3.3 Sales Policy

Sales policy refers to all measures that apply the sales strategy in the design of the sales system, the sales partnerships, the actual sales operations and the distribution of goods [41–43]. As part of crisis management, the sales responsibilities need to be

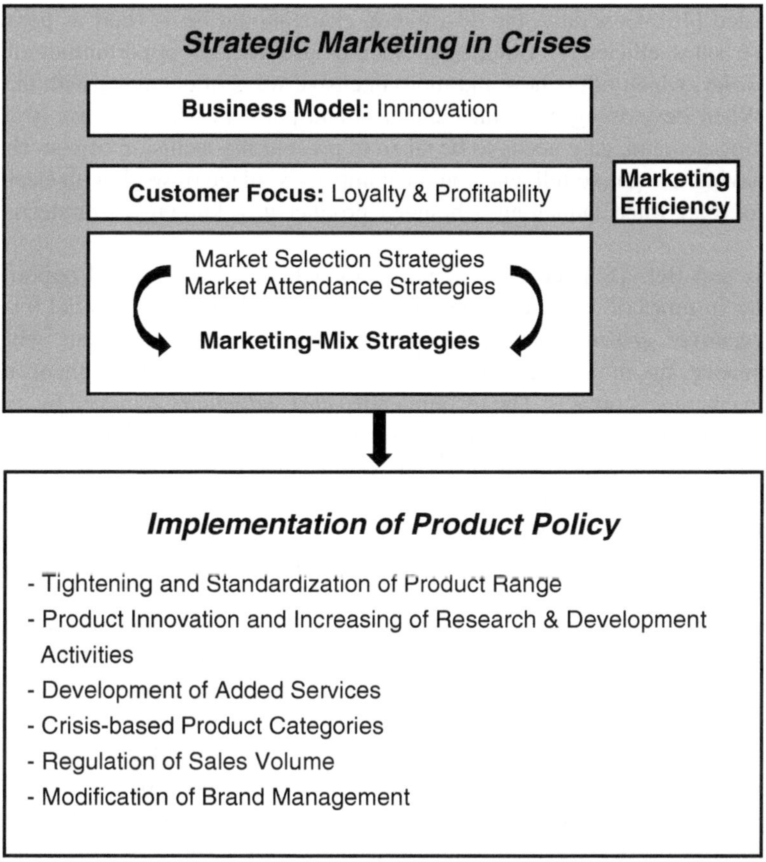

Fig. 7 Product policy in crises

subjected to a critical review with due consideration for the given circumstances (own vs. external sales operations; direct vs. indirect sales; single or multi-level sales; and single or multi-channel sales). This applies to the systems of cooperation, power or contractual relations with sales partners or key clients, the way of engaging with clients (personal contacts) and the logistical terms in the distribution of the goods to the client. Current literature offers a complete bundle of pragmatic recommendations that are particularly relevant for the various policy areas of crises [8, 24, 28, 30, 45, 46, 54–56]. In line with product policy, the basic assumption is that a difficult situation in the market will demand certain measures to maintain or increase turnover and improve the sales margin, while coping with substantial cost pressures.

One empirically proven lever for increased market share, revenue and profitability is the ***number and quality of sales channels***. In times of crisis, the used channels should be checked to see whether they should be developed or indeed eliminated [46]. Generally, the distribution chain should be as short as possible to increase sales efficiency. Companies should also use all opportunities afforded ***direct sales***, which offers new and more intensive ways of engaging with the client [45]. When developing new sales channels and partners while a crisis is actively impacting demand, care needs to be taken to prevent the inclusion of new channels from having a negative influence on the company's image or work with established partners (e.g. when launching a branded product in a known discounter market) [28].

Belz and Belz [57] see three perspectives with relevant practical opportunities for sales in times of crisis: concentration, saving or rebuilding. Potential for ***short-term turnover growth*** can be had by reinforcing work with clients with high performance, the more careful selling of key services and the investment of more time on customer contacts. The recommended ***cost reductions*** achieved by reducing commissions for sales professionals, reducing expenses or reducing the size of the sales force as a whole should be taken with a pinch of salt, as these options can come into a serious conflict with the aim of increasing turnover and revenue. Sales innovation, more professional sales practice, restructuring in sales or improved customer relations management are meant to generate ***new potential for future sales***.

Simon [45] mentions a set of emergency responses in sales and field sales, all of which have a similar thrust in common. In line with Faulhaber and Grabow [8], he also considers sales as the critical bottleneck factor in turnover crisis, which deserves priority attention from management. The first recommendation concerns the ***intensification of sales management*** and the careful reduction of inefficient sales activities [45]: "*savings in sales during a crisis means savings made at the wrong end. This does not, however, refer to targeted efficiencies*". Another option for reinforcing turnover lies in direct sales and in ***increasing the core sales time***. The relevant process improvements in this respect can focus, e.g. on the use of not fully employed internal sales capacities to provide support and relief to field sales. Internal sales personnel can even be developed to become fully qualified field sales professionals [56]. For added impulses, companies can consider poaching effective sales personnel from competitors. Room for improvement is also seen in more ***professional contact and route planning***, which includes the more sophisticated management of customer contacts and concentration on the more profitable categories of clients. In addition to increasing deliveries to existing customers, ***neighbouring customer segments*** can be developed by building up new regional, industry, or price-related segments. ***Monetary incentives*** are particularly relevant in sales, which should be aimed at actual profit and not simple turnover and should be designed to be fully compatible with any planned changes in pricing. Considering the question of ***knowledge transfer in sales***, established top professionals can attain an important role in training up the remainder of the sales team. Many times, opportunities for ***cross-selling*** are left unused. This also applies feeds into considerations of improved customer loyalty and creates new footholds for other ***sales partnerships*** that can lead to a complementary expansion of the product range.

The turnaround concept of Faulhaber and Grabow [8] distinguishes between ***six turnaround levers in sales*** :

- Realigning the sales strategy (refocusing on selected customer/product groups in a balanced mix, concentrating on successful regions and improving order sizes and make-up).
- Redesigning sales processes (reducing the number of contacts with less productive clients [30], administrative relief for field sales and restructuring internal sales).
- Training the sales team (improving key account management, job rotation for field staff and reinforcing tele-selling).
- Intensifying sales management (opt. with new executive personnel).
- Modifying incentive systems (reducing the fixed and expanding the variable components).
- Optimizing the sales IT systems (streamlining interfaces, using efficient CRM systems).

In his discussion of the implications of company crises for sales, Carter [55] emphasizes the ***person of the sales director*** and his or her duties. He applies a three-phase model for the development of crisis solutions to distinguish between the various functions and methods of relevance for sales executives. The primary responsibility lies in an intensive collection of the available facts and sales data, which provides a basis for developing different scenarios for sales during the crisis. Sales executives should be included in the crisis management team, depending on the nature of the crisis and the presence of the relevant competences, or indeed to be put in charge of it. They can be the troubleshooters and motivators who keep the sales force empowered and capable. Suitable remuneration rules that reward appropriate risk management and prevent too much risk-taking can achieve a focused risk response. At the same time, unprofitable transactions and general sales overhead need to be reduced. Crisis management should provide suitable software solutions to support ongoing sales. For the greater involvement of the client, customer advisory boards are proposed as suitable options. Generally speaking, crises are seen as opportunities for the evolution of sales and the sales or general company culture.

It is striking that the aspect of sales-related ***goods distribution*** is not really covered by current literature. It could be considered part of the general primacy of efficiency in the sales response to a crisis. The priority objects for a crisis-oriented sales strategy are covered by Fig. 8.

3.4 Pricing Policy

The price of goods and services gets a particular political relevance in times of crisis [39]. Frequently, price adjustments are the only decisions made by sales for a market-side response to the crisis. The reason lies in the simple flexibility of ***pricing***

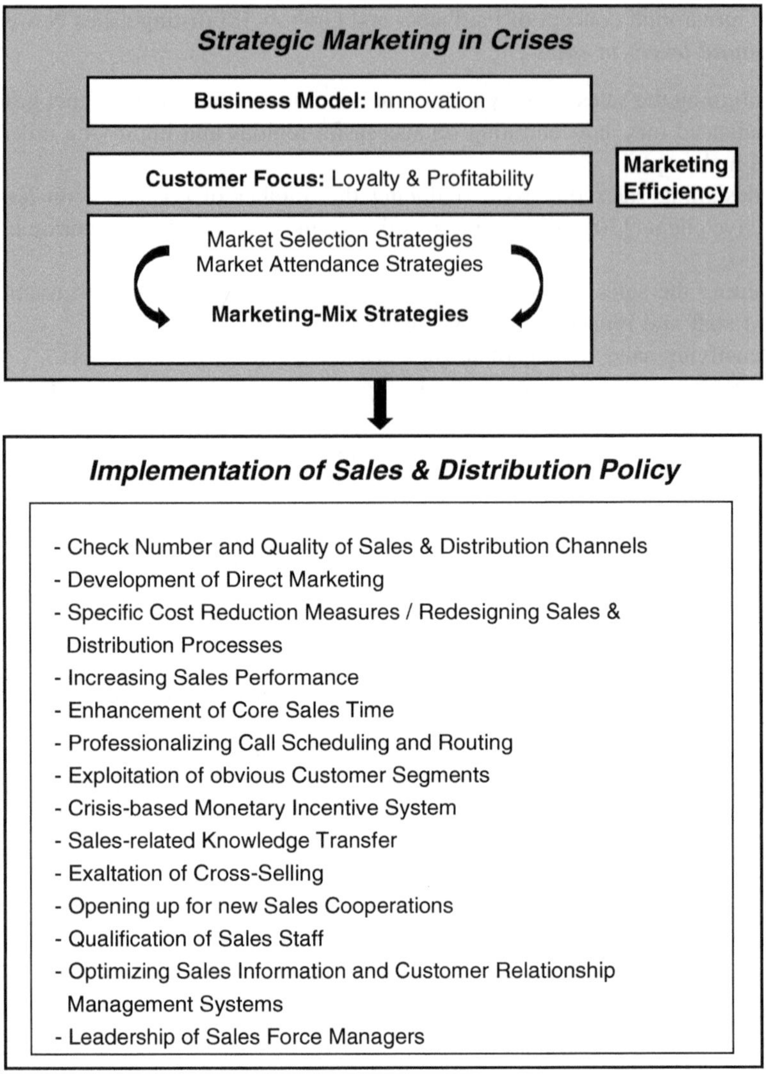

Fig. 8 Sales and distribution policy in crises

and the immediate impact it can have [41]. However, this also comes with the flip side of the great difficulty of taking back pricing decisions once made. With its simultaneous impact on the volume and value of sales, pricing has an important role to play for the commercial performance and profitability of any company [8, 56]. Relevant factors informing pricing policy include the elasticity of the consumer-side response to prices, the behaviour of consumers (e.g. interest in prices, reference prices, certain thresholds) or the nature of the market. The objects to be considered in the decision include the general positioning of the price (premium, general or discount markets), the product lifecycle and certain differentiation aspects (e.g. by region, by time, by target groups) [41]. Prices can be calculated by considering costs, demand or the competition, although the most common approach, also in crises, is a combination of methods that covers all perspectives and seems centred on the question of value for the customer [8, 58]. To introduce and enforce certain prices, other aspects of the terms and conditions also need to be considered, such as discounts, terms of payment or financing options. Companies should also remember that other perspectives of their policy also affect their pricing decisions [41–43, 56].

Empirical studies have produced different, sometimes contradictory findings about the effect of pricing policy. There is, however, a general consensus that prices are frequently adjusted during crises that often have a massive impact on the profitability and image of a company [28, 51]. Marasovic et al. suggest ***three options for crisis pricing*** [28] in the pursuit of increasing demand: maintaining the old price while increasing product quality; reducing the price while keeping quality constant, or a combined reduction of price and quality. Other feasible options like an increased price [56] combined with more variety in product quality or a reduction of supply are not addressed. Announcing higher prices for luxury goods can not only lead to improved revenues, but even a short-term spike in sales, as was seen when Daimler-Benz increased the list prices for its cars in the wake of the oil crisis of the 1970s [40]. Like most proponents of using pricing policy as a tool in crisis management, Marasovic et al. emphasize ***indirect pricing*** via changes to terms and conditions, which are easier to revert once the crisis has been overcome. In this respect, Simon emphasizes the need for nonlinear pricing and price bundling, that is in essence, offering volume discounts [45]. ***Maintaining the established prices*** while offering greater benefits for the customer is generally proposed as a preferred option.

Grappling with the developments in the wake of the 2008 financial crisis, Elste [59] bases his ***recommendations for price management*** in times of crisis on the signalling and demand-influencing effects of pricing policy and on its function in terms of creating value and recovering costs. More than half of the industries affected by the crisis decided to lower their prices by substantial margins in order to counter the often double-digit drop in demand [39, 59]. As product quality typically remained unchanged, companies later found it exceedingly difficult to revert back to their original prices and were confronted with questions and criticism about the honesty of their pricing policy. Any ***price reduction*** in a crisis should therefore only be introduced in tandem with a reduction in the service offered and only as long as a negative impact on margins is unlikely when price–value relations remain unchanged [30, 59, 60]. Companies could also consider reducing prices while keeping quality at the original level if they are subsidized by public support, as in the case of too-big-too-fail corporations or industries. Defending the price is one of the core objectives of crisis management. Ruinous ***price wars*** need to be avoided at all costs [8, 39, 45, 51, 59]. This needs acceptance of the price leadership of the dominant provider and active and early information for competitors, as far as possible under normal antitrust legislation. ***Volume*** is therefore the preferred lever in times of crisis or in general cases of excess capacities. This can also mean offering material discounts that are easier to retract again than actual price discounts [45]. In the case of major ***cost spikes***, e.g. caused by a sudden upsurge in the cost of raw materials, the extent to which such costs can be handed down to the customer should be checked, e.g. in the form of direct material cost surcharges. The pricing and terms and conditions should be ***transparent*** and easy to understand, while also following the old adage that the customer should not notice any price increases (staying "under the radar", e.g. by imposing fees for formerly free services), while discounts should be made public [45]. A crisis therefore also represents a great opportunity to simplify ***pricing policies*** that have become overly complex and inconsistent over time and return to fair pricing [37] in terms of the target clientele. Many companies can accept losses of up to 31 % because of the uneconomically minute orders of small-scale clients. Losing such clients because of crisis-driven price increases can indeed represent an advantage for these companies. The principle should be that discounts can become greater, the greater the contribution margin of the client [8]. In terms of ***price-driven sales management***, the key is to close old pricing loopholes (e.g. hidden reimbursements or credit) and spread awareness in the sales team for the increasing importance of the price–value argument [8, 59]. Furthermore, pricing policy and its effects can be tracked via dedicated ***indicators*** to immediately recognize any relevant deviations. It would seem particularly relevant here to quantify and monitor the price-turnover function [45]. Established rules for calculating prices are, essentially, not challenged by crises; however, they need to be enforced more consistently and with more sense for the given circumstances in the pursuit of more professional price management in crises [58, 59]. Figure 9 reveals the most important objects of crisis-oriented pricing.

3.5 Communication Policy

Communication policy obtains a particularly important role in crisis marketing, as it can override, support or even undermine all other activities. It can be understood, resembling strategic marketing, as an integrative framework bringing the marketing mix for use in the crisis together. At the same time, it has the potential to break up that mix and fuel the escalation of the crisis. There have been many instances of poor communication in a crisis becoming the actual engine of the crisis [61]. Speaking generally, ***communication policy*** means the targeted influencing of the attitudes and behaviour of certain target groups by means of the planned communication of information. With a specific set of circumstances to work with, it deals with the right selection and creation of communication messages, instruments and audiences. The target audience defined by means of specific criteria should be the basis for marketing communication, focusing primarily on the customer and his specific traits [41, 42]. Typically, more general corporate communications are also considered part of the marketing function, which expands the scope of such communication activities to all stakeholders of the company (Public Relations und

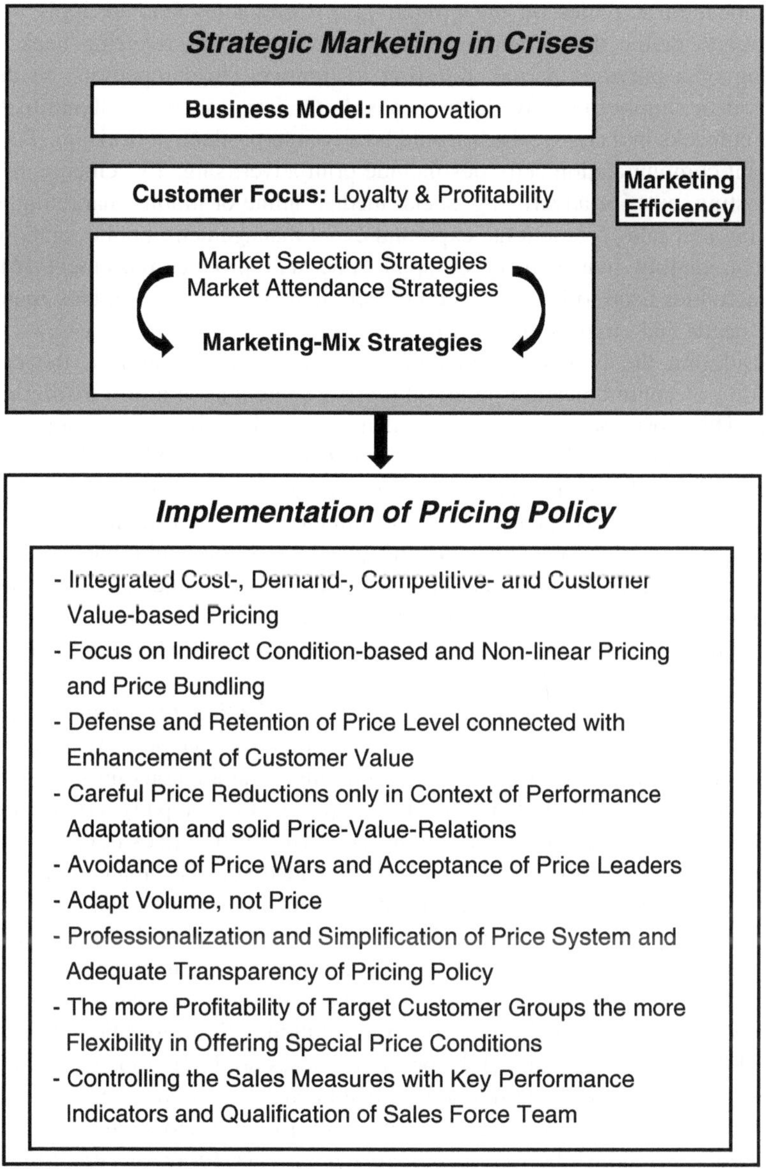

Fig. 9 Pricing policy in crises

Corporate Identity policy). PR activities that are appropriate for the given situation and the targeted stakeholders are particularly crucial under the conditions of corporate crisis [62, 63]. The traditional planning process begins by defining the goals and target groups for communication and then defines the communication budget, plans and designs the media activities, and conducts and monitors the communication activities on the ground [42]. Companies tend to apply specific heuristics to define their communication budgets, such as referring back to the spending of a previous period, referring to turnover, to competitors, to defined objectives or simply to the available means. In the last case, this will automatically lead to cutbacks in a crisis, which would be a counterproductive decision. The most important communication activities include print advertising, TV, cinema, or radio broadcasting, exterior advertising, social media, online or mobile marketing, direct marketing and sales promotions, expo and event management, public relations and corporate identity management or sponsoring and product placement [41–43]. These activities need to be scrutinized closely in a crisis to see that they match the unique needs and circumstances in it.

Considering the two trends in current literature on the subject, the creative structuring of communication policy in a crisis context should also ***distinguish its objects***. The most common area covered in publications is crisis communication, that is, communicating the crisis to the relevant stakeholders. What is less common, but increasingly frequent since the economic crises of 2008 and later, is the exploration of the consequences of crisis situations for the established aspects of marketing communication. This concerns questions of the marketing mix in terms of communication *in a crisis*. The focus here lies on communication with clients to recover or strengthen demand. Both perspectives need to be seen in close connection with each other and should be treated as an integral whole by marking in a crisis.

Crisis communication in the sense of ***communicating the crisis*** is typically recognized as a crucial element of corporate crisis management. It concerns the problem-centric response to current and often highly ambiguous information [64], using internal and external communication to protect the company's reputation and sustain loyalty as a means of overcoming the crisis itself. Crises have the habit of attracting public attention, and a response is called for. The manner in which a company responds to this can have a negative effect on its entire industry [65], since such crises, depending on their severity, are interesting and, from a journalistic standpoint, very attractive topics. They concern questions of uncertainty, individual stories, of blame and of sheer curiosity, although the latter concerns mostly the start and growth of the crisis and not its end or resolution. Crisis resolution attracts only a fifth of the media attention that the previous phases do. For companies, this means that bad news need to be cushioned with immediate, proactive, constant, unambiguous, truthful and empathetic communication [61, 64]. Höbel and Hofmann [8, 39, 66] both distinguish between ***six factors for successful crisis communication***: time (speed before comprehensiveness), personnel (a qualified communication team), trust (empathy vs. fear), complexity (reduction and conflict resolution), cost (investments into crisis communication pay of) and expertise (crisis manual as a pool of knowledge). These relate closely to the characteristic traits of different types of crisis (accident, disaster, image, personnel, change management or mergers, hijacking or blackmail, compliance, information, legal change, product, finance, international crisis) that represent the situational context for the communication going on in them.

Crisis communication normally takes places under severe time pressure, which can be mitigated by introducing ***preparatory measures*** before a crisis hits [61, 64, 66–69]:

- Forming trust in the relevant public.
- Establishing media contacts.
- Developing a crisis plan (manual) with instructions concerning the methods, contents and responsibilities in communication.
- Coordinating human resources.
- Providing a crisis centre with the right equipment.
- Telecommunication, outbound services and call centres.
- Providing images and video material for broadcasters.
- Developing darksites (easily upscalable crisis websites).
- Establishing a protected internal information platform.
- Using a alarm software for potentially relevant parties.
- Crisis training (TV and media training, phone and interaction training, simulations, stress and self-management).

In view of the fact that crises are increasingly becoming a recurring fact for businesses everywhere, preparing for crisis communication can be approached as part of corporate development and change management.

After the crisis has materialized, the aim is to take proactive action to maintain control over the discourse. It is essential to avoid ambiguous messages and to ensure that everybody involved speaks with "one voice". Corporate communications, corporate wording and corporate behaviour need to be fully aligned, especially in times of crisis. Honesty is another prime directive in order to maintain one's position of trust among stakeholders after the crisis has hit. The frequency and level of detail in communication depend on the urgency of the crisis [39]. Crisis communication needs to take account of its inherent dynamics, which makes it necessary to constantly rethink the impacts [64, 67]. Of particular relevance in terms of the inherent dynamics and potential loose-cannon quality of crisis communication are ***social and online media***. Companies in crisis are faced with an active subset of the wider public with an opinion-shifting impact even on traditional media outlets [61, 66, 67, 70]. Many crisis communicators also tend to neglect their ***internal audience***, that is, the employees and their representatives or the executives not included in the crisis team [71]. These also need to be involved from the very onset of the crisis; employees need to be kept informed about what the press knows, and they can act as important multipliers for the crisis messages in their presence on social media. They should know the official language and understand that any expression of their private opinions can be taken up by media and should therefore be marked as such [8, 61, 66]. The worst-case scenario is the exclusion of the workforce from crisis communication, forcing them to revert to online media or the social network rumour mill for their own information.

When ***assessing and developing*** concepts for crisis communication or engaging in communication after a crisis has been overcome, official practice guidelines can be of use for companies. One of the most quoted sources is the US National Center for Food Protection and Defense (NCFPD) and its publication of best-practice guidelines for risk and crisis communication with the following key concepts [72, 73]:

- Planning ahead (crisis guidelines) for an immediate response.
- Establishing a broad network for crisis communication.
- Accepting insecurity (tell what you know).
- Forming external partnerships during the crisis.
- Being aware of the public perception of the crisis.
- Open, honest and credible communication.
- Knowing and responding to the needs of the media, always being available.
- Expressing compassion and empathy.
- Providing ideas for the people affected by the crisis to protect themselves.
- Always updating the crisis plans.
- Accepting and considering the cultural differences at work (stakeholder-specific communication).

Similar recommendations for risk or crisis communication can be found in the practice tips of specialist consultants [e.g. 74].

Communication in a crisis considers the crisis only a specific phase in which marketing communication policy is enacted. The all too frequent and unthinking cutting back of communication budgets in crisis is generally rejected by professionals in the field [35, 75]. A number of empirical studies have revealed the benefits of the anticyclical ***maintenance or even extension of the communication budget*** as against its reduction as a deceptive source for cost savings [28]. Economic crises actually make the success of communication efforts more likely as the consumption of media increases, while the cost of media services falls [60]. In such crises, advertising should ***emphasize customer benefits*** to account for the stricter rational choices made by consumers in these periods. On common recommendation and actually frequent observation is the reallocation of advertising efforts to cheaper and more innovative ***online media*** [28]. Specifically, this can mean reinforcing one's social network presence, engaging in more search engine work, integrating online and offline media, starting more customer and sales-oriented email communication, or redesigning corporate web pages [35, 51, 75]. In a crisis, reinforcing marketing activities on social networks is a particularly promising option when it comes to attracting new clients. It is economical, has the potential for viral marketing and represents a global, immediate and emotionally charged access to the client [75]. It is also important to ***involve the sales team*** in communication during a crisis. The direct communication efforts of sales professionals remain the most important and most cost-efficient options for marketing. They should be supported by intensive sales promotions [28]. As in crisis communication proper, communication in a crisis also has the creation and protection of trust as its primary purpose, although with a much stronger focus on the customer. Consumers that are more cautious with their spending money and more distrustful should be reached with complementary ***public relations activities***, instead of traditional advertising [28, 53, 69]. Apaydin and Geçti [30] would favour the cost

Fig. 10 Communication policy: crisis communication and communication in a crisis

aspects in communication; when considering the practices of companies during crises, they see possible ***savings*** in reducing advertising spending, standardizing advertising, performance-oriented pay for advertising agencies and using low-cost communication channels. They too recognize the increasing importance of public relations and online activities, but they disagree with the sales-focused recommendations of Marasovic et al. in emphasizing the savings options in the field of quite costly sales promotions. Whether these should be increased or, indeed, reduced should depend on the specific sales constellation in each case. Each company needs to weigh up the cost impact of changes and their effects on revenue, as should be done for all communication efforts. Concerning the ***stronger focus on the customer*** in times of crisis, it serves to remember the commercial rationale of focusing sales activities on selling solutions for the customers' problems: "*offerings, information and communication must be based not on what the company wants to sell, but what customers may need to buy*" [51].

Figure 10 illustrates the most important aspects of communication policy in times of crisis.

4 Summary

Our investigation of a contingent approach to the crisis-specific management of different policy aspects in marketing was occasioned by the apparent rise in the number of crises experienced by companies in the real world, their relevance for corporate development and the surprising lack of a comprehensive analysis of crises as a phenomenon from the functional perspective of marketing. Crises were defined as highly dynamic, complex and competitively relevant "peak" events of these three qualities of the business environment at a specific point in time. They bring about a specific constellation of risks with often life-threatening impact for companies. An indicator of the relevance of such crises, not least in the German-speaking countries, is the constantly high number of insolvencies, which the general public only tends to notice in the rare instances of spectacular company collapses. Corporate crises are recurring events in the evolution of any company with a special constellation of internal and external conditions. They can be considered critical, but "normal" incidents.

Corporate crises are "home-grown" first and foremost. That means that the forces threatening the survival of a company, even if they come into effect as a result of external events, can always be traced back to poor management before or during the crisis. Current literature offers a vast array of categories of corporate crises to support a professional response to the phenomenon in business practice. Combined with the established triad of strategy, performance and liquidity crises, contingent crises (e.g. emergencies, health risks caused by a product) are described as exogenous shocks that come into effect on all levels, i.e. at all stages of their development and that can force the company's hand when their impact is made worse by the forces of complexity, dynamics or competitive relevance. The creative aspects of crisis management come to the fore when a distinction is made between strategic and operational crisis management. Strategic crisis management concerns the prevention of crises, the response to strategic crises and the management of the strategic aspects of operational crisis management, which is tied in with the strategic perspective, but often has to take precedence because of the urgency and dynamics of an acute crisis. From the strategic or operational approach to crisis management, one can arrive at more specific functional categories in the traditional management areas such as procurement, production, marketing, human resources or finance.

For the theoretical approach to marketing in a crisis context, we drew attention to the situational approaches of organizational theory or the fit perspective of strategic management. In this sense, the situational, contingent properties of crises determine the structure and behaviour of marketing management, intentionally or unintentionally. The contingency approach for marketing in crises aims at achieving a "crisis-marketing fit" to form the frame of reference that the different policy areas in marketing use in the pursuit of that right fit.

The first, general creative area concerns strategic marketing, which is often a formative force for the company's general strategy or even a direct mirror image of it. Three areas for exploration and creative intervention were identified in strategic marketing: the business model as the nucleus of the corporate and marketing strategy, whose evolution reflects the need and ability to innovate in a time of crisis; the focus on the customer, which means involving and aligning marketing activities with the actual needs of the customer, retaining customers and maintaining profitability in customer transactions; and marketing (sub-) strategies as the functional expression of the corporate strategy. All three areas are subject to the overriding mission of protecting the efficiency of marketing and lead to concrete action plans in the various policy areas of the crisis-marketing mix.

A set of typical responses was identified for product policy, which includes brand management: Standardizing and streamlining the product range, redoubling the effort invested in innovation and R&D, offering additional services, considering the importance of crisis-specific product categories, fine-tuning the sales volumes or modifying the brand carefully. For sales policy, the proposed responses are more versatile again: they range from rethinking the number and quality of sales and distribution channels, focusing on direct sales, introducing crisis-specific incentive systems or development opportunities for the sales team to targeted economy measures in sales. Another list of possible responses could be proposed for pricing policy, such as holistic cost, demand, competition- and customer-oriented pricing, focusing on indirect price management via specific terms and conditions, or avoiding ruinous price wars. The last question of policy, that is, communication policy in the sense of *communicating the crisis*, has attracted most attention in the literature on crisis management. It helps to distinguish a second perspective, i.e. *communication in a crisis*. There are viable options for both perspectives, such as emphasizing the importance of pre-planned crisis communication (communicating the crisis), communicating the added value for the customer to potential consumers (communicating in a crisis), or engaging in trust-inspiring activities as an important area for both types of communication.

Crises are a normal part of any company's development, but their specific nature also warrants a specific marketing response, adjusted to match the conditions of each individual set of circumstances. This general statement, however, also reveals the inherent weakness of a contingency-driven response to crisis management, as we need to ask whether it is possible to propose any general recommendations as each and every crisis comes from a different cause and takes a different course. Nonetheless, this is a basic problem that faces all disciplines of management theory and constitutes a dilemma that business economics, as a creativity-oriented science, has always grappled with and will still face in future. The systems and policy recommendations made here are therefore only intended as an offer of support, a possible means of simplifying the practical response to this phenomenon of business which no company can hope to avoid at some point in its story. Turning the argument around for a more positive outlook, a confrontation with marketing policy for crisis situations is, by needs, a great occasion to do marketing as it should always be done.

Review Questions (Yes or No)

1. Are all corporate crises preceded by a specific constellation of risks?
2. Can all corporate crises be traced back to internal causes?
3. Should spontaneous crises be treated as a type of crises different from strategy, performance or liquidity crises?
4. Is the management of a strategic crisis the same as the management of the strategic aspects of a crisis?
5. Is the contingent nature of the situational response only relevant for structural questions?
6. Is marketing efficiency the key paradigm in strategic marketing?
7. Does an expansion of the product range as a product policy decision typically help alleviate a crisis?
8. Should direct sales normally be intensified in times of crisis?
9. Is supply management to be preferred to pricing management in sales crises?
10. Is crisis communication the same as communication *in a crisis*?

Answers to the Review Questions

1. Yes / 2.1
2. Yes / 2.2
3. No / 2.2
4. No / 2.3
5. No / 2.4
6. No / 3.1
7. No / 3.2
8. Yes / 3.3
9. Yes / 3.4
10. No / 3.5

References

1. Krystek U, Lentz M (2014) Unternehmenskrisen: Beschreibung, Ursachen, Verlauf und Wirkungen überlebenskritischer Prozesse in Unternehmen. In: Thießen A (ed) Handbuch Krisenmanagement, 2nd edn. Springer VS, Wiesbaden
2. Krystek U (1987) Unternehmungskrisen. Beschreibung, Vermeidung und Bewältigung überlebenskritischer Prozesse in Unternehmungen. Gabler, Wiesbaden
3. Witte E (1981) Die Unternehmenskrise – Anfang vom Ende oder Neubeginn? In: Bratschitsch R, Schnellinger W (eds) Unternehmenskrisen – Ursachen, Frühwarnung, Bewältigung. C.E. Poeschel Verlag, Stuttgart
4. Roselieb F, Dreher M (2008) Reden und Handeln sind Gold – Wie erfahrene Krisenmanager kritische Situationen meistern. In: Roselieb F, Dreher M (eds) Krisenmanagement in der Praxis. Von erfolgreichen Krisenmanagern lernen. Erich Schmidt Verlag, Berlin
5. Bundesamt für Statistik Neuchatel (2015) Konkursverfahren und Betreibungshandlungen, 2011–2014
6. Statistisches Bundesamt Wiesbaden (2015) Insolvenzen von Unternehmen und übrigen Schuldnern. https://www.destatis.de/DE/ZahlenFakten/GesamtwirtschaftUmwelt/Unternehmen Handwerk/Insolvenzen/Tabellen/UnternehmenSchuldner.html. Accessed 23 July 2015
7. Kreditschutzverband von 1870 Wien (2015) KSV 1870 Insolvenzstatistik 2014 – Pressekonferenz vom 07.01.2015. https://www.ksv.at/pressekonferenz. Accessed 23 July 2015
8. Faulhaber P, Grabow H-J (2009) Turnaround Management in der Praxis. Umbruchphasen nutzen – neue Stärken entwickeln. Campus Verlag, Frankfurt a.M
9. Hauschildt J (2006) Entwicklungslinien der empirischen Krisenforschung Unternehmen. In: Hutzschenreuter T, Griess-Nega T (eds) Krisenmanagement. Grundlagen – Strategien – Instrumente. Gabler, Wiesbaden
10. Krystek U (2006) Krisenarten und Krisenursachen. In: Hutzschenreuter T, Griess-Nega T (eds) Krisenmanagement. Grundlagen – Strategien – Instrumente. Gabler, Wiesbaden
11. Hauschildt J (1983) Aus Schaden klug. Manager Magazin 13(10):142–152
12. Hauschildt J, Grape C, Schindler M (2005) Typologien von Unternehmenskrisen im Wandel. In: Schriftenreihe aus den Instituten für Betriebswirtschaftslehre der Universität Kiel (CAU), 588, Wiesbaden
13. Müller R (1986) Krisenmanagement in der Unternehmung: Vorgehen, Massnahmen und Organisation, 2nd edn. Lang, Frankfurt a.M
14. Töpfer A (2009) Krisenmanagement. Verlauf. Bewältigung und Prävention von Krisen. WiSt 4:180–187
15. Töpfer A (2014) Die Managementperspektive im Krisenmanagement – Welche Rolle spielt das Management bei der Krisenbewältigung? In: Thießen A (ed) Handbuch Krisenmanagement, 2nd edn. Springer VS, Wiesbaden
16. Baier-Fuchs A (2014) In der Krise ist vor der Krise – wie man durch systematische Vorbereitung Krisen eindämmt und Katastrophen verhindert. In: Thießen A (ed) Handbuch Krisenmanagement, 2nd edn. Springer VS, Wiesbaden
17. Bea FX, Haas J (2013) Strategisches Management, 6th edn. UVK Verlagsgesellschaft, Konstanz
18. Ernst FA (1997) Die Integration von unternehmens- und personenbezogenen Lebenszyklen - Eine Konzeptualisierung unter besonderer Berücksichtigung des Unternehmenslebenszyklus. Dissertation, Universität St.Gallen Strategisches Management, 6th edn. UVK Verlagsgesellschaft, Konstanz
19. Gomez P, Zimmermann T (1993) Unternehmensorganisation: Profile, Dynamik, Methodik. Campus Verlag, Frankfurt a.M
20. Pümpin C, Wunderlin C (2005) Unternehmensentwicklung. Haupt Verlag, Bern
21. Bleicher K (2011) Das Konzept Integriertes Management, 8th edn. Campus Verlag, Frankfurt a.M

22. Greiner LE (1994) Evolution and revolution as organizations grow. In: Mainiero L, Tromley C (eds) Developing managerial skills in organizational behavior: exercises, cases, and readings, 2nd edn. Prentice Hall, Englewood Cliffs
23. Becker J (2013) Strategisches Management, 10th edn. Verlag Franz Vahlen, München
24. Dubrovski D (2014) The role of marketing restructuring in a company crisis. Int J Econ Pract Theories 4(5):658–666
25. Bea FX, Göbel E (2010) Organisation. Theorie und Gestaltung, 4th edn. Lucius & Lucius, Stuttgart
26. Lawrence PR, Lorsch JW (1986) Organization and environment: managing differentiation and integration. Harvard Business Review Press, Boston
27. Chandler AD (1962) Strategy and structure. Chapters in the history of the american industrial enterprise. The MIT Press, Cambridge
28. Marasović I, Crnjak Karanović B, Dragnić D (2011) The impact of crisis on marketing strategy components—An overview of empirical evidence. In: Křižman Pavlović D, Benazić D (eds) 22nd CROMAR congress: marketing challenges in new economy. Juraj Dobrila University of Pula, Pula
29. Rosier ER (2011) Marketing strategy in a turbulent environment. J Strateg Mark 19/5:413–419
30. Apaydin F, Geçti F (2011) How do enterprises change their marketing strategies in an economic crisis? A field study on turkish enterprises. Int J GSTF Bus Rev 1(1):243–248
31. Jurse M, Korez Vide R (2010) Strategic thinking as a requisite management tool for managing international marketing in turbulent times. In: University of Zagreb (ed) International conference—An enterprise Odyssey: from crisis to prosperity—Challenges for Government and Business. University of Zagreb, Opatija
32. Shama A (1978) Management & consumers in an era of stagflation. J Mark 42(3):43–52
33. Lombriser PA, Ablanalp R (2010) Strategisches Management. Visionen entwickeln – Erfolgspotenziale aufbauen – Strategien umsetzen, 5th edn. Versus, Zürich
34. Ansoff HI (1976) Managing surprise and discontinuity – strategic response to weak signals. Zeitschrift für betriebswirtschaftliche Forschung 28(28):129–152
35. Mitran PC, Bebeşelea M (2012) About the crisis marketing and the crisis of marketing. Econ Manage Fin Markets 7(4):660–665
36. Makovec Brenčič M, Pfaifar G, Rašković M (2012) Managing in a time of crisis: marketing, HRM and innovation. J Bus Ind Mark 27(6):436–446
37. Mansoor D, Jalal A (2011) The global business crisis and customer behavior: Kingdom of Bahrein as a case study. Int J Bus Manage 6(1):104–115
38. Naidoo V (2010) Firm survival through crisis: the influence of orientation, marketing innovation and business strategy. Ind Mark Manage 39:1311–1320
39. Loidl F, Sabo J, Wührer G (2010) (eds) Absatzkrisen anders bewältigen. Konsequenzen für Marketing und Management. Linde Verlag, Wien
40. Fischer D (2010) Krisen und Krisenbewältigung bei der Daimler-Benz AG, Dissertation, Friedrich-Alexander-Universität Erlangen-Nürnberg
41. Meffert H, Burmann C, Kirchgeorg M (2015) Marketing. Grundlagen marktorientierter Unternehmensführung. Konzepte – Instrumente - Praxisbeispiele, 12th edn. Springer Gabler, Wiesbaden
42. Homburg C (2014) Marketingmanagement. Strategie – Instrumente – Umsetzung – Unternehmensführung, 5th edn. Springer Gabler, Wiesbaden
43. Kotler P, Armstrong G, Wong V, Saunders J (2011) Grundlagen des Marketing, 5th edn. Pearson Studium, München
44. Schneider W (2013) Strategisches Marketing. Von der Planung zum strategischen Profil. Oldenbourg Verlag, München
45. Simon H (2009) Sofortmassnahmen gegen die Krise. Wege für Ihr Unternehmen. Campus Verlag, Frankfurt a.M
46. Müller J (2013) Turnaround. Ein Leitfaden für Manager und Verwaltunsräte. Verlag Neue Zürcher Zeitung, Zürich

47. Dugal SS, Morbey GK (1995) Revisiting corporate R&D spending during a recession. Res Technol Manage 38(4):23–27
48. Vodopivec R (2012) Influence of political globalization and global crisis on traditional marketing management theory and practice. Procedia Soc Behav Sci 44:330–340
49. Elliott D, Harris K, Baron S (2005) Crisis management and services marketing. J Serv Mark 19(5):336–344
50. Esch F-R, Rempel JE (2006) Krisenmanagement für und durch Marken. In: Hutzschenreuter T, Griess-Nega T (eds) Krisenmanagement. Grundlagen – Strategien – Instrumente. Gabler, Wiesbaden
51. Grundey D (2009) Branding strategies during economic crisis: avoiding the erosion. Econ Sociol 2(2):9–22
52. Dahlĕn M, Lange F (2006) A disaster is contagious: how a brand in crisis affects other brands. J Advertising Res 12:388–397
53. Greyser SA (2009) Corporate brand reputation and brand crisis management. Manage Decis 47(4):590–602
54. Binckebanck L (ed.) (2011) Verkaufen nach der Krise. Vertriebliche Erfolgspotenziale der Zukunft nutzen – Strategien und Tipps aus Forschung, Beratung und Praxis. Gabler Verlag, Wiesbaden
55. Carter T (1997) Crisis management for sales force managers. J Prof Serv Mark 87–103
56. Laker M (2006) Profitable Umsatzsteigerung in der Krise. In: Hutzschenreuter T, Griess-Nega T (eds) Krisenmanagement. Grundlagen – Strategien – Instrumente. Gabler, Wiesbaden
57. Belz C, Belz O (2011) Verkauf in schwierigen Zeiten. In: Binckebanck L (ed) Verkaufen nach der Krise. Vertriebliche Erfolgspotenziale der Zukunft nutzen – Strategien und Tipps aus Forschung, Beratung und Praxis. Gabler Verlag, Wiesbaden http://www.bfs.admin.ch/bfs/portal/de/index/themen/06/02/blank/key/02/betreibungen.html. Accessed 23 July 2015
58. Fasnacht M (2010) Auswirkungen der Krise auf das Marketing. Zeitschrift für Controlling & Management, Sonderheft 1:24–25
59. Elste R (2011) Hat die Krise alle Grundsätze der Preissetzung ausgehebelt? Empfehlungen für den Vertrieb. In: Binckebanck L (ed) Verkaufen nach der Krise. Vertriebliche Erfolgspotenziale der Zukunft nutzen – Strategien und Tipps aus Forschung, Beratung und Praxis. Gabler Verlag, Wiesbaden
60. Rhodes D, Stelter D (2010) Nach der Krise ist vor dem Aufschwung. Wie Unternehmen die stagnierende Wirtschaft für Überholmanöver nutzen. FinanzBuch Verlag, München
61. Riecken M (2014) Erfolgskritische Faktoren der angewandten Krisenkommunikation. In: Thießen A (ed) Handbuch Krisenmanagement, 2nd edn. Springer VS, Wiesbaden
62. Coombs WT (2007) Protecting organization reputations during a crisis: The development and application of situational crisis communication theory. Corp Reputation Rev 10(3):163–176
63. Stephens KK, Malone PC, Bailey CM (2005) Communicating with stakeholders during a crisis. J Bus Commun 42(4):390–419
64. Merten K (2014) Krise, Krisenmanagement und Krisenkommunikation. In: Thießen A (ed) Handbuch Krisenmanagement, 2nd edn. Springer VS, Wiesbaden
65. Sarstedt M (2009) Reputation management in times of crisis. J Brand Manage 16:499–503
66. Höbel P, Hofmann T (2014) Krisenkommunikation, 2nd edn. UVK Verlagsgesellschaft, Konstanz
67. Raupp J (2014) Krisenkommunikation und Media Relations. In: Thießen A (ed) Handbuch Krisenmanagement, 2nd edn. Springer VS, Wiesbaden
68. Garth AJ (2008) Krisenmanagement und Kommunikation. Das Wort ist Schwert – die Wahrheit Schild. Gabler, Wiesbaden
69. Sojung K (2013) Does corporate advertising work in a crisis? An examination of inoculation theory. J Mark Commun 19(4):293–305
70. Perry DC, Taylor M, Doerfel ML (2003) Internet-based communication in crisis management. Manage Commun Quart 17(2):206–232

71. Heide M (2014) Internal crisis communication and management. In: Thießen A (ed) Handbuch Krisenmanagement, 2nd edn. Springer VS, Wiesbaden
72. Veil SR, Husted RA (2012) Best practices as an assessment for crisis communication. J Commun Manage 16(2):131–145
73. Seeger MW (2006) Best practices in crisis communication: an expert panel process. J Appl Commun Res 34(3):232–244
74. Ferrante P (2010) Risc & crisis communication. Essential skills for today's SH&E professional. Prof Safety 06:38–45
75. Tomše T, Snoj B (2009) Marketing communication on social networks—Solution in times of crisis. Marketing 45(2):131–138

Understanding Digital Marketing—Basics and Actions

Teresa Piñeiro-Otero and Xabier Martínez-Rolán

Abstract This chapter provides a technical outline of the basics of online marketing. The outline includes an introduction to digital marketing and strategic planning and development. Our contribution offers theoretical and practical insights relative to this growing marketing area, with information on the main areas for which online marketing is particularly suited: (1) the E-commerce section explores different business models and what techniques are used for their development; (2) Web Search Marketing focuses on SEO and SEM, as well as in keyword selection for optimisation; (3) E-mail Marketing offers interesting content to develop a successful newsletter; and (4) Social Media Marketing addresses planning and the most important tools used to maximise communication through social media. In a nutshell, this chapter offers an overview of digital marketing and its strategies for an active and effective Web presence.

1 Introduction

In *The Third Wave*, Toffler [1] predicted the demarketisation of postindustrial societies. Three decades later, the end of marketing is still not in sight, even though, as Kotler [2] already suggested, marketing had to rethink its foundations to adapt to Third Wave societies and individuals.

In 1999, Schutz and Holbrook [3] referred to the *tragedy of the commons* to stress the low efficiency of market strategies due to overuse and reiteration of strategies and tools. In the last decade of the twentieth century, organisations were

T. Piñeiro-Otero (✉)
Faculty of Communication Sciencies, University of A Coruña, Campus de Elviña s/n, 15071 A Coruña, Spain
e-mail: teresa.pineiro@udc.es

X. Martínez-Rolán
Faculty of Communication and Social Sciences, University of Vigo, Campus de Pontevedra s/n, 36005 Vigo, Spain

C. Machado and J.P. Davim (eds.), *MBA*,
Management and Industrial Engineering, DOI 10.1007/978-3-319-28281-7_2

forced to use more resources to reach dwindling audiences. Consumer resistance to marketing actions highlighted a deep crisis.

This context of communicational saturation in which an increasing number of brands compete for the attention and loyalty of audiences forced traditional marketing perspectives and concepts to change. During this process, the most relevant change was the power shift from marketers to consumers.

The consumer perspective has been present in the definition of marketing since the 1960s [4] even if it was only in recent decades that it took centre stage in any marketing strategy. This approach has led to the incorporation of aspects such as consumer satisfaction, market orientation or consumer value in marketing management [5].

However, many marketers still think in terms of product, place, promotion and price, McCarthy's variables or 4Ps model [6], which does not leave any role to consumers. This production-focused marketing paradigm was later challenged by Lauterborn's user-centred models [7]. The 4Ps of the marketing mix yield to the 4Cs that turn product into customer solution, price into cost to the customer, place into convenience and promotion into communication. This is a new perspective for operational marketing that will be of special relevance for the online world.

In recent years, further steps have been taken—marketing does not only focus exclusively on consumers but also tries to bring different audiences to organisations. This new approach was defined by the American Marketing Association as an activity, set of institutions and processes for creating, communicating, delivering and exchanging offerings that have value for customers, clients, partners and society at large [8].

The Internet boom for organisations and the daily life of different audiences brought about a deep transformation of marketing, its tools and strategies [9].

Although initially organisations understood the Internet as a new channel to increase their presence, they soon started looking for ways to maximise its different platforms and services. At present, online marketing communications are an essential part of operational marketing from the point of view of becoming, in themselves, a specific marketing line: digital marketing.

2 What Is Digital Marketing?

The first approaches to digital marketing defined it as a projection of conventional marketing, its tools and strategies, on Internet. However, the particularities of the digital world and its appropriation for marketing have fostered the development of channels, formats and languages that have led to tools and strategies that are unthinkable offline.

Today, rather than a subtype of conventional marketing, digital marketing has become a new phenomenon that brings together customisation and mass distribution to accomplish marketing goals. Technological convergence and the multiplication of devices have led to an opening up of the ways in which we thinking about marketing in Internet and have pushed the boundaries towards a new concept of digital marketing—user-centred, more measurable, ubiquitous and interactive.

The development digital marketing strategies offer much potential for brands and organisations. Some of them are as follows:

- Branding. Platforms and 2.0 services are a great opportunity to build a brand image on the Web due to their scope, presence and constant updates.
- Completeness. The possibilities to disseminate information through links offer consumers the chance to approach the organisation in a wider and customised way.
- Usability–functionality. Web 2.0 offers simple and user-friendly platforms for all in order to improve user experience and allow for their activities.
- Interactivity. In the context in which organisations try to forge long-term relationships with their audiences, Internet offers the possibility of having a conversation and therefore of generating a positive experience with the brand. Such interactivity can be basic, as product assessment, or become an all-encompassing experience.
- Visual communication. In line with visual thinking, digital marketing offers marketers different image- and video-based tools. This is an attractive way of reaching audiences that can lead to greater engagement.
- Relevant advertising. Easy segmentation and customisation of advertising in Internet maximise the output. Besides, free from the limitations of other media, this environment has allowed for more attractive advertising.
- Community connections. Internet is a unique opportunity to connect organisations with their audiences and users among themselves. This connectivity can improve their experience and enhance the relationship with the product, brand or organisation.
- Virality. The essence of Internet as a Web of interconnected nodes makes exponential expansion of any content possible. Taking the model of WOM (word of mouth) communication, viral communication becomes more relevant due to connectivity, instantness and shareability of online platforms that enhance the dissemination of content.
- Measuring output. Online platforms rank first in the availability of follow-up options and the possibility to assess output.

In any case, to make the best of all these possibilities, organisations must ensure that their Internet presence or their presence on their different 2.0 channels follows a strategy with concrete goals, in line with their brand or organisational image. Being on the Web without proper planning can not only mean a lost opportunity in terms of resources and potential, but also it can indeed have a negative impact on the organisation, as the audience, their needs and perceptions regarding the organisation are unknown.

3 Digital Marketing Plan

A digital marketing plan is a strategic document that takes the current situation of a particular organisation to set some midterm goals and to determine the strategy and means to accomplish them. This document also describes the responsibilities, the time frame and control tools for monitoring.

The aims of a digital marketing plan include discussing organisations and their environment. Likewise, it needs to be a roadmap of how to manage the organisation's marketing strategy, so that resources are properly allocated. The plan also helps control and evaluate output and tackle any potential deviation from the organisation's expected outcomes. In this line, a marketing plan becomes a flexible document that must be adapted to the situation of the company and that must feed into the results obtained by each of the actions developed, especially in the digital arena.

The development of an online marketing plan is similar to a conventional one in its structure, but it also includes some variations at an operational level. In fact, this means not only that some specific strategies and tools are to be developed, but also that the volubility of the digital media and its capacity for immediate measurement force organisations to develop a strategy that can be revised in the short term, as well as the iterative process between action and control.

An online marketing plan is a document in line with the company's strategic plan that sets goals of an activity in the digital environment, as well as the what, how, when, who and why (6 Ws) of Web presence.

There are different options regarding the structure of a marketing plan. We advocate for a four-phase structure as this is simple and clear to plan for any action's strategy (Fig. 1).

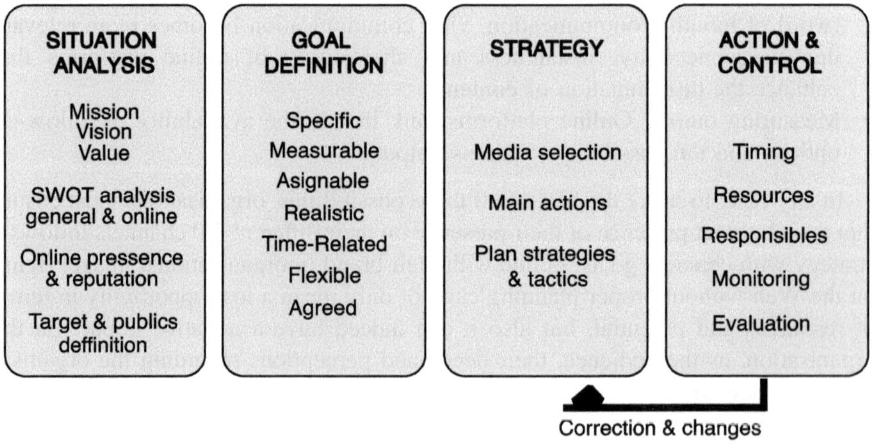

Fig. 1 Marketing plan. *Source* Prepared by the authors

3.1 Situational Analysis

One of the most relevant parts of any strategic document is the definition of the baseline situation of the organisation and its environment. This analysis allows for the development of an action plan that is reality based and shall therefore minimise the risks in the development of a strategy.

The starting point of any situational analysis is to define the mission, vision and values of the organisation. This definition helps guide any marketing action.

- The mission defines the chore of the company, its essence. This definition usually includes the work or activity of the organisation, reference to its audience, business models and the singularity or differential factor of the organisation. The mission responds to questions such as who am I?, where do I come from?, what do I do?, whom do I address?, what is my field of operations? and what is my competitive edge?
- The vision must describe the future goals of the organisation in a short and concise way. These goals must be realistic and achievable in order to motivate all stakeholders in their achievement. The organisation's vision answers questions such as where am I going to?, what do I long to be? and where do I want to be?
- The values are the principles that guide the business culture and that the organisation—must—fulfil in all its activities.

Once we have defined these three essential aspects from a communicational and strategic perspective, the next step is the development of an internal–external analysis of the organisation.

An internal analysis is relevant to determine weaknesses and strengths, and it must address aspects related to production, commerce, organisational and financial issues, as well as the attitude of management.

For an external analysis, the goal is to determine the threats and opportunities, thus considering the specific environment (products, clients, competitors and suppliers), as well as the general environment, namely the ecological, technological, economic, legal, political or socio-cultural constrains.

Both analyses lead to a SWOT matrix.

A digital marketing plan, apart from being a study of the organisation and its environment, must analyse its online presence and position. This study shall help determine a concrete digital strategy and shall help compare its results with the initial situation. In this sense, the following must be developed:

- Monitoring of the organisation's keywords, competition and sector (some useful tools are Mention, Google Alerts, Hootsuite).
- Web positioning assessment of the organisation and its platforms (e.g. Google search or Alexa ranking).
- Social network evaluation. Presence, activity, influence, etc. (here tools, such as Klout, PeerIndex and Kreed, become relevant).

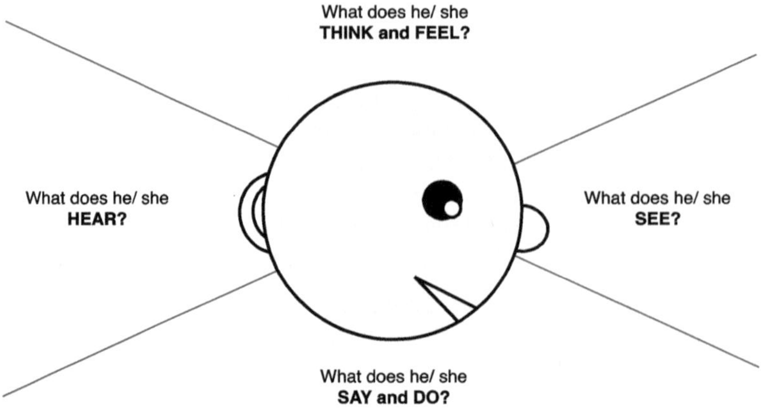

Fig. 2 Empathy map. *Source* XPLANE [10]

- Competition benchmarking and main influencers in the sector present on digital media.
- Specific SWOT.

In order to complete the situational analysis, audiences must be defined, with special attention to the target group, without neglecting the remaining stakeholders. This description must also focus on the presence and activities of these audiences in the digital world: active presence in platforms and services, access devices, usage times, contents of interest, main activities, etc.

The explanation of audiences must be very detailed in trying to adjust the proposal both to the real needs of those users and to their aspirations and frustrations (Fig. 2).

This phase helps determine the organisation and its situation in the environment in which it operates and, more particularly, in the digital world. From this information, the potential and gaps of the organisation can be identified to design a future strategy.

3.2 Goal Setting

Setting goals is one of the key phases in any marketing plan. Goal setting offers an idea of where the organisation is going and it enables adjustments to the marketing strategy.

Goal setting must follow SMART criteria [11]

- Specific: simple and easy to understand.
- Measurable: they can be measured through any kind of quantitative or qualitative unit.

- Assignable: they can be assigned and implemented by a member of the company.
- Realistic: with accessible resources, goals can be achieved in a realistic framework.
- Time-related: they must have a time frame to achieve them.

In the specific field of online marketing, such goals must also be flexible to adapt to the changes in the company and Web evolution and reached through a consensus. Many of the actions of digital marketing are horizontal—for example actions on social media—and involve different departments in the company; therefore, consensus is key.

In an online marketing plan, the most usual types of goals are as follows:

- Scope goals.
- Activity goals.
- Conversion goals.
- Loyalty goals.

Once these goals are set, they must be implemented through key performance indicators. These indicators enable measurement and monitoring.

3.3 *Planning a Strategy*

In order to accomplish the planned goals, a specific strategy must be defined. A strategy is the implemented scheme to achieve such goals. This strategy materialises in activities that are the ways planned to achieve such goals.

An online marketing strategy is based on the model of a marketing funnel suggested by Strong [12] as a development of the AIDA model (Awareness, Interest, Desire, Action). The transformation of marketing in recent decades and the particularities of the digital world have allowed for a revision of this marketing model so that it captures the conversion strategy, as well as the loyalty strategy, key in the online world.

There are several proposals, such as that by Rogers [13], who advocates that, apart from Awareness, Consideration, Preference and Action (an update of the AIDA model phases in the current context), two new states must be incorporated: loyalty and advocacy. Like Strong's model, each phase means a higher level of commitment, and therefore, loyalty and advocacy are at the bottom of the model (Fig. 3).

However, in the online world, loyalty and support for the brand can lead to change consumers into prescribers, thus increasing the scope of the organisation's action, which could lead to an expansion of its consumer base.

In the phase of determining a strategy, means and actions must also be defined. The translation of conventional marketing to the Web, as well as the multiplication of 2.0 platforms and services, has stressed the complexity of the new media reality.

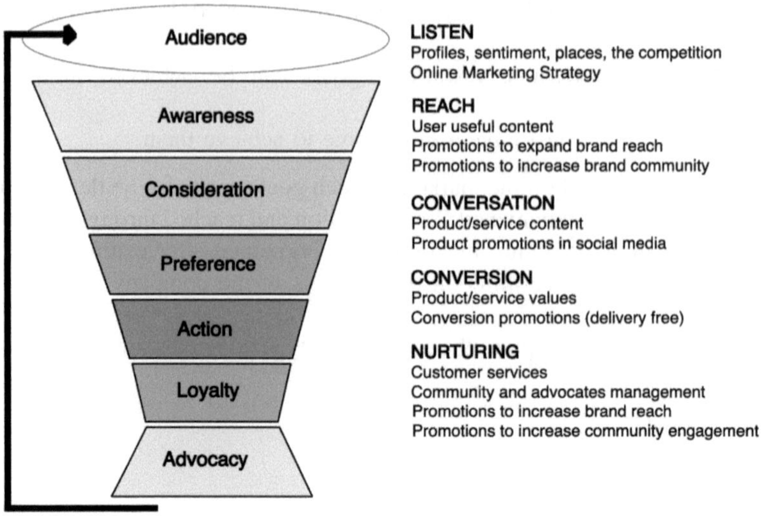

Fig. 3 Marketing funnel review. *Source* Prepared by the authors based on Rogers [13]

While the classification of paid, earned and owned media has been a constant in marketing, in the past differences in the three categories were clearly defined. Most marketing initiatives focused on paid media, while owned and earned media were used to reinforce and amplify advertising messages. Loss in efficacy in advertising and the appearance of the social Web has led to rethinking this model by blurring the lines between different media types.

Their convergence is taking root in digital channels, rapidly moving from one type of media to another. Companies must know and combine the three types of media in order to ensure greater effectiveness in the building process of their own audiences.

In the context in which users are more critical with the organisations and in which peer assessment is given more credit than brands, earned media become key against paid or owned media. Many funds can be allocated to advertisement, but one cannot force consumers to assess their experience or recommend a product or service in particular.

Any comment about a particular brand on Internet can have a measurable impact in terms of scope; therefore, the goal of organisations is to have a significant relationship with their users so that they convert into consumers and, on a higher level of commitment, prescribers (Fig. 4).

Despite the increase in relevance of earned media, the three types must be used in a coordinated fashion to maximise the efficiency of a digital marketing strategy.

- *Owned media* Corporate channels such as websites, social network profiles and mobile apps. These channels are unique, and organisations have total control over their content within the limits of each platform. Owned media offer avenues for most content distributed by the organisation, and thus, they become the

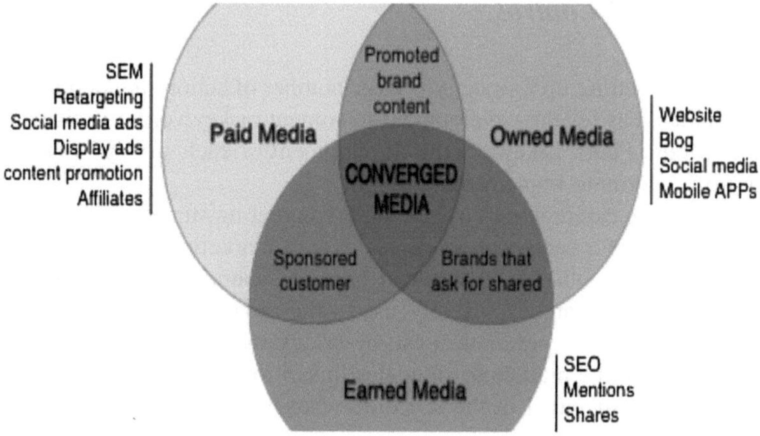

Fig. 4 Convergence media. *Source* Prepared by the authors based on Lieb and Owyang [14]

backbone of the digital strategy of a brand. However, due to the fact that they are biased communication, they do not always have the push or scope needed.

- *Earned media* Content about the brand developed by external users, for free. Earned media is essentially online word of mouth: mentions, shares, reposts, reviews, recommendations, product reviews on specialised Webs or assessments on online shops or specific social platforms (e.g. Ciao! or TripAdvisor). Peer recommendation is essential in the current context, as it is perceived as genuine and unbiased, as brands lack control over them. One of the most effective tactics to get public support for brands is to combine a good organic positioning in browsers (SEO) with a good content strategy.
- *Paid media* These are spaces or content that the brand had to pay for. They tend to be used to foster or increase the scope of the messages and initiatives of the brand in owned or earned media, as well as to improve their output. Apart from advertisement display and advertisement in social media, earned media includes retargeting, Pay Per Click, promoted tweets, paid reviews in specialised blogs or recommendations by influencers on social media (they should be identified as such). Good management of paid media fosters content for earned media, as well as increasing traffic in owned media. Such an increase in the Web can lead to sales of a product or services.

Each of these media offers their own advantages and scenarios; therefore, using them complementary can lead to accomplishing the set goals. All of them contribute to the development of a digital marketing strategy, although each brand must analyse what media is more appropriate for them and on which the return on investment (understood in its widest sense) is higher.

Despite the fact that actions in digital marketing have changed to adapt to new platforms and users, some of the most relevant in terms of use and results are e-commerce, Web search marketing, e-mail marketing and social media marketing. Their features and strategies are developed in the following sections.

3.4 Action and Control

In the process of setting up a strategy, a large number of actions are defined, and the possible theme areas for the development of content or keywords are listed (e.g. in the Social Media Plan). Likewise, the development of each strategy requires time planning to programme implementation.

In this sense, an essential part of any digital marketing strategy is the scheduling of tasks and timing for each of them. In the case of marketing strategies based on a limited number of tools—or example a branding campaign where a Facebook profile and newsletter are exclusively used—the content of each of the actions (publication, sending the newsletter) can be briefly mentioned in the scheduled.

Likewise, for online marketing strategies linked to traditional marketing, offline actions have to be included in the schedule whenever they are interdependent on online tasks.

Scheduling means assigning tasks in the digital marketing strategy, with clear indication of who is responsible for each action. Therefore, all departments in the organisation should have a copy of the marketing plan, especially those departments and professionals involved in its proper development.

In the action phase, the cost of different actions planned must also be considered. In this estimate, both technical and operational costs related to the implementation and monitoring of the online strategy are to be clearly included, especially those related to websites or online shops.

In the case of other 2.0 tools, despite the fact that most of them are free for the development of marketing strategies, the cost of the professional developing and monitoring the strategy must be included or, otherwise, training of staff and purchase of specific management tools for the strategy, its monitoring and output assessment.

Constant feedback and quantitative data of interactions allow for constant control of the planned strategy's evolution. However, dates must be set to assess the different actions and tools in detail, looking at the specific features of each of them. This can be a quarterly evaluation for SEO, monthly for the corporate Web or weekly for some strategies on social media.

These actions can be reviewed and corrected depending on the output of each individual action; however, the value of the development of the complete strategy must be periodically assessed. Generally speaking, 4–6-month intervals are recommended for such assessments, always taking into account the total duration of the digital strategy and the type of techniques used.

The conclusions to review the strategic document shall be drawn from this process. This document has to be updated to maximise the efficiency of the following implementation phase. At the end of each phase, the process shall be repeated.

4 Social Media Marketing

Social media are a great opportunity to establish significant relationships and create ways of social interaction defined through dynamic exchanges between their members. Social media is booming in terms of the number and variety of platforms and users.

Thus, one can find audiovisual platforms such as YouTube, Vimeo and SoundCloud; image platforms such as Flikr, Picassa, Pinterest or Instagram; general social networks such as Facebook, Twitter, Google+ or specialised ones such as LinkedIn; news or bookmark aggregators such as Digg or Delicious; blogs; and wikis, etc., a vast digital arena where they become the new Web winners.

Although the notion of a social network is not new, it has reached new heights, thanks to Web penetration and connectivity. Social networks have developed through platforms that show different types of functions, but common features. These features aim at creating a community by connecting users, who can interact, discuss, offer insights or knowledge. Technology in the case of these platforms also needs to be flexible and conducive to an exchange of information. This means that free web standards; and modular architectures that lead to complex but efficient applications are usually favoured [15].

In essence, a social network manager is a service that allows individuals to create a public or semi-public profile within an enclosed system, to articulate a list of other users they share connections with, and to use that list of users, as well as other nodes in the system. The nature and nomenclature of such connections may vary from one site to the other [16].

Such platforms become content containers as they grant space and tools for a user who plays an increasingly more active role. In this sense, brands and organisations should be part of the social conversation and use the interactive channels to listen to their users, who share their perspectives, insights and knowledge, or generate value through content development.

4.1 Prosumers: Paradigm of an Active User

Internet penetration and the democratisation of some information and communication technologies have favoured the creation of a World Wide Web with constant information flow. In this mesh of social digital relationships, each user or node can become a content producer.

The breaking down of the classical division between sender and receiver has favoured the appearance of the prosumer (acronym for consumer and producer), an active user who not only accesses and uses content, but also produces and disseminates it, thanks to the appropriation of 2.0 tools. Despite the fact that the idea that a user is both a content producer and consumer had already been presented by Toffler [1], this was seen as a utopia until the advent of Web 2.0. Users produce more content on a daily basis than any other classical senders such as corporations and media, thus becoming the undisputable pillar of the Web.

On the social Web, Internet users have access to a number of free platforms with simple and user-friendly interfaces that allow them to create their own content, make it available for thousands of users and get feedback in just a few minutes.

The democratisation of such collaborative platforms paved the way for an alternative production system based on crowd wisdom that calls for a rethinking of the marketing strategies by organisations. More and more users look for precision, relevance, power and reciprocity in marketing, the actual bases for concurrence marketing. Precision and relevance refer to the agreement with the consumer in both targeting and messaging. On the other hand, power and reciprocity are linked to cooperation with the consumer and product design or to marketing implementation [17].

With customisation, concurrence marketing becomes an opportunity for the establishment of long-term relationships with audiences. This is a new scenario that has led to the transformation of social interactions, information access and use, a scenario that has forced brands to stop bombarding consumers to have a conversation with them instead.

Organisations have started looking for touch points to their different Internet audiences. These touch points are seen as spaces for interaction whereby brands can trigger strategies to add value and transform their messages in an attractive experience for the user. Marketing is more effective when it is consumers who look for brands rather than the other way around.

4.2 The Role of Community Managers

In a new environment of interactions in which brands and their audiences share space and conversations on an equal footing, a new professional profile has emerged: that of community managers.

A CM (community manager) is a professional in charge of the social platforms of brands or public persons, so that they become the link between the organisation/person and its community. The roles of a CM are community based, as they are responsible for building and managing a community, as well as for content production and activation.

The tasks of a CM include listening to a social audience and identifying opinion leaders, an action that may help them optimise management of their community while they gather value-added information for the strategic management of the organisation. In this way, the CM becomes the voice of the company for external audiences and the voice of the social community for marketing directors.

Increasing specialisation of the sector has led to the specialisation as community managers, who, as Baston [18] highlighted, in agencies of digital marketing and large corporations, shares space with more specific social profiles such as:

- Social Media Manager: they coordinate the community managers.
- Social Media Strategist: they interpret the reports and designs a strategy for social media.

- Social Media Analyst: they have a technical profile as data analysts and interpret the metrics to draft reports.
- Social Media Developer: this is a hyperspecialisation of a programmer who knows the social APIs and designs project code for social media (e.g. competitions or specific landing pages).
- Social Media Public Relations: they foster content outside their own channels.
- Social Media SEO: they apply SEO techniques to social media, especially for campaigns of online reputation.
- Content Curator: they select and organise relevant information.

4.3 Social Media Plan

A Social Media Plan (SMP) is the master document that guides an organisation's presence on social media, and it covers all aspects to be considered when setting up, maintaining and integrating social networks in the organisation's digital marketing strategies. Therefore, any Social Media Plan must be in line with the marketing plan.

A Social Media Plan is a living document that needs close control due to the liveliness of the virtual world in which it is implemented.

This document usually follows a relatively stable structure including goal setting, types of audiences, platforms, strategies and tools, and output measurement. With output measurement, the whole process begins anew (Fig. 5).

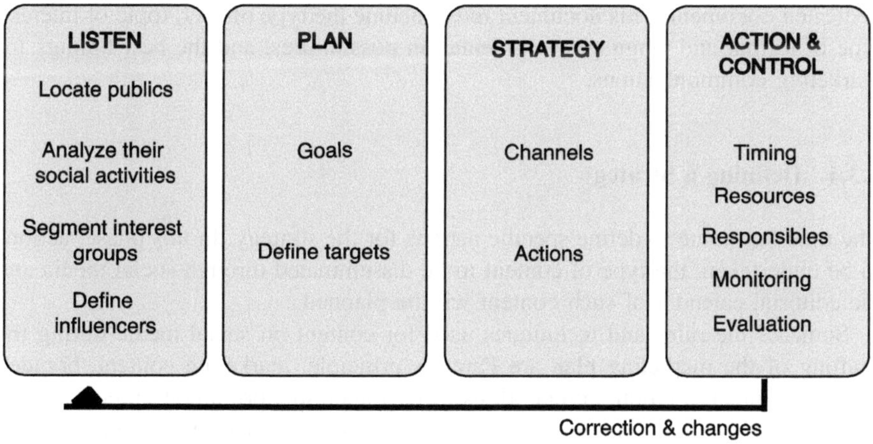

Fig. 5 Social Media Plan. *Source* Prepared by the authors

4.3.1 Goals

Apart from following SMART criteria when they are discussed, the goals of a marketing plan must cover three different nonexclusive targets:

- Visibility: making the brand popular (e.g. more Facebook hits).
- Sales: to increase sales or leads through a specific platform (e.g. to turn visits to the site into actual sales through twitter links to products of an online shop).
- Loyalty: to preserve an audience that has already been attracted (e.g. through a media-based customer support service).

4.3.2 Audience

The objectives must be targeted at a specific audience. The more you know the organisation knows its target audience, the more possibilities it has of fostering efficient communication and forging a significant relationship. In the case of social media, most of these data are given by the users themselves when they create their profiles. Data are further defined by actions and interactions among users in social platforms and, generally speaking, on the Web. It is also common to use statistics portals to select more specific audiences.

4.3.3 Channels

Goal setting and a definition of target audiences give insights into the channels to be used for our social media marketing strategy. To materialise this decision in an efficient way, each of the suggested platforms should be schematically analysed in a dedicated document. This document must include the type of user, topic of interest, type of format and languages, segmentation possibilities, and the best timings for marketing communications.

4.3.4 Defining a Strategy

The time has come to define specific actions for the strategy. In this phase, actions to be undertaken, the type of content to be disseminated through social media and the editorial calendar of such content will be planned.

Some of the rules and techniques used for content on social media during the drafting of the marketing plan are Pareto's principle, marketing content, branded content, content curation, competitions or customer support, among other.

Content Marketing: these are actions to create and disseminate relevant and useful content to raise interest in the audience and attract them, instead of interrupting them so that they buy products and services [19].

Despite the fact that content marketing is not a new phenomenon—see the monthly magazines of airlines—it has boomed with Web 2.0. Through their own media, brands disseminate useful and quality content to attract and retain qualified traffic. Content must be of quality and useful, must have the brand personality and be multiformat and segmented.

Furthermore, content marketing tries to change users into sneezers, expanding the scope of the brand beyond its own channels.

Formats to implement this strategy are varied: from more traditional formats such as white papers, e-books or specialised magazines to other types of formats specific to the digital world such as podcasts or videos, or other types of content with higher virality such as infographics or memes.

Content Curation: this is the process of collecting, selecting, organising and adapting the relevant information on certain topics or trends on the Web to be published in an attractive and significant way.

Content curation allows the company to be constantly updated and know new Web features, especially in terms of its audiences, while it saves time in the process of content creation.

Having a content curator can help brands design their content action plan and indicate possible developing lines depending on the topics of their audiences in the social conversation or who they follow on the Web. Tools such as Google Alerts or Google Trends and Mention, RSS readers such as Feedly or bookmarks such as Scoop or Delicious enhance identification and selection of topics of interest.

Customer Support: customers come to brands through social media looking for direct, useful, fast and effective contact. Thus, managing customer support through 2.0 platforms completes the customer experience with the brand.

Social networks allow for direct and instant contact with the company, fulfilling the concrete need of the customer at a critical point that, if successfully handled, can generate loyalty from the user, or undo a conversion otherwise. In both cases, the user can comment online about their experiences with the company and enlarge the scope of their experience.

Competitions: they are one of the best tools to promote an organisation's presence on social media, especially during product launch. In order to properly manage competitions, the following needs to be considered:

- Competition strategy: it must be adapted to the possibilities of each social media. Using 2.0 specific services for competitions such as Cool Tabs or Offerpop for Facebook is encouraged.
- Regulations: they have to be clear and concise and provide a detailed explanation of the operations, dates, participation, award, etc.
- Prize: it must be attractive and therefore encourage participation. If a product or service by the company is given as prize, it can not only reduce the competition cost, but also help convert a user into a customer and afterwards into a fan.

4.3.5 Measuring Outcomes

Assessing the actions taken on social media is vital to measure success and check whether the goals set have been accomplished. As there is a large variety of measurable items for social media, KPIs must be concretely described and adjusted to the goals in order to develop a successful Social Media Plan; otherwise, there is a risk of going into vanity metrics that do not measure real performance [20].

A KPI can cross-check different metrics to measure an objective. For example, in order to know whether popularity on Facebook has increased, one can use the number of hits or interactions with the brand in a particular period, compared to competitors or a previous period.

Metrics usually revolve around the following factors:

- Audience: community volume (e.g. Facebook fans, Twitter followers). This is one of the most basic metrics. Quantitatively, they can indicate the evolution of a community, but they do not offer qualitative data about it.
- Scope: this is related to the size of the community, although it really measures direct amplification. It can be measured on Facebook through the number of shares, on Twitter with retweets, etc.
- Engagement: this is one of the most highly valued metrics; it measures the degree of engagement of the audience with the brand. It helps detect stakeholders and real fans.
- Influence: this measures the repercussion that is generated in the audience, so that it can be seen as part of engagement. One of the most widely used KPIs to measure influence is the Klout Index.
- Interaction: this is a complex metrics because it involves engagement and brand perception. On Facebook, for example, this is measured by PTAT (people talking about this), a metrics that counts the users that in some way have interacted with the brand.

4.3.6 Pivoting

Once the actions implemented are evaluated and the accomplishment of goals has been checked, some conclusions must be drawn so that the organisation can pivot, integrating precise changes in its marketing strategy on social media.

4.4 Social Media Advertisement

Most social networks show business models based on advertising. If we take for granted that a social network manager is a massive database, with large amounts of qualitative data from its users, using them allows brands for microsegmentation of their advertising actions.

Facebook is the king of segmentation. Its advertising platform, Facebook Ads, allows for a delimitation of the target audience of each ad depending on location, age, sex, languages and even interests and behaviours. Facebook is the social network with more data on its users. The basic data filled in by the user alone in the process of registration on Facebook includes name, e-mail, sex and age of the new user.

Facebook advertising formats are inserted either on the sidebar on the right-hand side of the platform itself or on the users own timeline (Web and mobile), as well as on the logout page. These adverts must to be identified as such, while their goal is to foster as much interaction with users as possible. They thus try to increase engagement, lead users to a website or special offers, get more likes on pages, app downloads... etc.

Twitter, on the other hand, has less information on users; therefore, its segmentation capacity is lower. However, through its advertisement platform, Twitter Ads, they have launched some highly efficient publicity formats. This is the case of Twitter cards, a format focusing on promoted tweets to generate tweet engagement, website clicks or conversations, app installs or app engagements, followers or leads on Twitter, to name a few.

Other advertisement formats on Twitter are promoted accounts, which are placed on the profile recommendation section, or sponsored trending topics on the list of topics at a particular point in time.

5 E-mail Marketing

E-mail marketing is an online marketing technique that uses e-mail to send advertisements or commercial information. This is a communication tool used to attract new customers or make those that one already has loyal to the brand.

E-mail and Internet have gone hand in hand since the Web was created. The beginning of Internet dates back to 1969 (Arpanet at that time), while the first e-mail was sent two years later (1971). This first e-mail showed some basic features that have remained till the present: the use of "@" on the user name, as well as the fields "To", "Subject" and "Message".

In such a volatile environment, e-mail has been one of the Web tools that have best adapted to change, both in content and in scope and penetration. Therefore, e-mail marketing becomes one of the main tools in a digital strategy.

At present, e-mail is the first Internet service ahead of social media. In 2015, the number of e-mail accounts in the world was about 4.353 million users from which 205 billion e-mails were sent [21]. This volume of traffic includes legitimate e-mails and spam.

The term "spam" refers to those messages we do not request and we do not want or with an unknown sender, usually sent though mass mailing. Although spam can be used on other platforms and devices, for example SMS on mobile phones, e-mail is the most important channel for this practice.

The line between commercial information sent by e-mail and spam is, in many cases, a fine one that experts in e-mail marketing must properly identify in order to avoid making mistakes in their work.

The most common form of e-mail marketing is the newsletter: a publication that is distributed with a specific periodicity on an interesting topic for all recipients, called subscribers.

The complexity level of a newsletter shall depend on its goals and content; they can be simple with mainly plain text as the predominant feature or be enriched with images, graphs, adverts and/or hyperlinks.

The objectives of e-mail marketing can be multiple and varied; however, all of them could be grouped around four main goals:

- Diverting traffic to our website: be it the home page or any special section inside it.
- Promoting a special action: either promotion of new services, special discounts, sales, download of applications, etc. When the goal is to increase traffic or some Web-based special promotion, specific websites are usually created. Such pages are called landing pages.
- Cost savings: e-mail marketing supports order management and information to the customer regarding the status of such orders, as well as the provision of customer support services at a lower cost than other communication channels.
- Brand popularity and image: same as for other types of campaigns online, e-mail marketing is suited to generate popularity and brand image among consumers.

5.1 Advantages of E-mail Marketing

The strong penetration of e-mail in the current context becomes an important reason to include it any digital strategy, but there are also other important reasons to do so.

This is mass technology that instantly reaches everyone and whose use spans devices and screens (desktop, laptop, mobile telephones, tablets, etc.).

The system is direct and able to reach individuals in a scalable and targeted way. This is due to the fact that an e-mail can be sent to a single address or thousands, while content can target different types of audiences . In fact, the system's scalability does not hinder customisation; despite its capacity for mass mailing, it can be highly customised at a cost far lower than that of other types of campaigns.

This is also a multimedia channel that offers the possibility of sending a large amount of information as text, images (static or moving), sound or hyperlinks, in any combination.

As this is digital communication, its impact can be quantified through different metrics that allow for an evaluation of the output of each campaign.

5.2 Disadvantages of E-mail Marketing

Despite the many plus points of this marketing format, starting e-mail campaigns can be hindered due to some disadvantages inherent to this channel:

The most important enemy of professionals in the sector are antispam filters of e-mail managers that identify e-mail marketing as spam, thus reducing the effectiveness of campaigns.

Furthermore, explicit authorisation by the end-user to receive ads through mail is needed. Such actions are regulated by data protection acts and specific e-commerce laws.

Depending on the country and specific laws, sending unauthorised messages may result in very high fines for the companies sending them.

This need for prior authorisation to include the person on an e-mail database of a particular organisation has pushed e-marketing into the group "permission marketing" [22].

5.3 Legal Framework

The legal framework regulating e-mail marketing is important as it can inhibit some of its features.

On the one hand, legislation protects personal data in order to prevent the illicit transfer of databases. The relevant legislation forces the company to have a register—in some cases physical—of user data and makes the company responsible for the protection of such data.

On the other, each country develops laws to regulate mass mailing, a standard that influences commercial communications or advertising and those of transnational or relational character.

Despite the fact that the legal framework can vary from one country to the other, there are some common elements:

Mailing must have explicit authorisation by the recipient. This authorisation must be prior to the inclusion of the person in the mailing list.

The fact that the message is linked to advertising must be explicit, as well as the identification of the sender on the e-mail, the subject and the heading of the message.

In the cases of offers, competitions and promotional games, they must be identified as such and be clear and explicit about the conditions and participation on them.

In some countries, the advertising message has to be identified with the word "advertisement" or abbreviation, as well as with a valid postal address for the company.

Simple procedures for the user to withdraw consent are a must.

In the case of multinational companies, the relevant legislation is that of the country where the company is based and not that of the recipient.

5.4 The Heart of E-mail Marketing: Subscribers

The success of a marketing campaign lies precisely in target management; therefore, a good strategy to attract and retain subscribers is needed.

The recipient database must be meticulously kept and segmented. The essential question here is quality over quantity; therefore, purchasing user databases is discouraged—unless this is unavoidable, or unless their usefulness has been proven.

Best practices in e-mail marketing stress the need to generate and maintain the organisation's own database. The reason is simple: it is easier to get back a client than to get a new one.

Getting subscribers is a slow but steady task, although it can be sped up through online and traditional channels. Registration boxes can be created on a website, either as pop-ups or in another Web section where registration is required to request information (e.g. a budget) or in exchange for exclusive content or downloads.

Generally, easy forms must be used, where the compulsory fields include name and/or e-mail, although the number of fields in the form may be increased depending on the value of the treat one is offering the user in exchange for their registration.

Another common instance is e-commerce. When a user registers on a particular online shop and accepts the use and service terms, they are automatically included in the e-mail marketing database.

Regarding traditional channels, the compilation of e-mails and information related to the users can be done through competitions, on-street promotion, postcards, loyalty cards or formal information requests through e-mail, phone or postal mail. Traditional channels demand, however, the digitalisation of the data to be included in a subscriber list.

Keeping subscribers is a task that needs special attention and that must be implemented while new users are added. In this process, brands must be especially careful regarding content and the form e-mail communications take. Content must be interesting for the user, giving priority to quality instead of periodicity.

In a nutshell, it is about adding value to commercial communications with complementary information (comparisons, advice, features, etc.) useful for the consumer.

5.5 Newsletters—Some Key Aspects

Four key aspects are to be considered for effective newsletter design:

- Database. In order to have good content segmentation, the organisation needs to know the subscriber base of their newsletter as much as possible. This information will allow for specific profiling depending on demographics, geographical data, interests, etc., thus increasing the efficiency of communications.
- The content of the newsletter. The content must be useful and interesting, and the weight of commercial information must be properly balanced. Subscribers

value content that adds value or is useful to them. In fact, newsletters are one of the key tools of content marketing.
- The field "from". In the context of communication saturation, the customer will look at the field "from" as a filter to recognise the origin and dismiss—or not—the mail.
- The field "subject". Same as with a slogan or tagline, the field "subject" must stand out above the mass of incoming e-mails. Likewise, this field must fulfil the criteria of usefulness and interest for the user and, above all, honesty about the content they will find in the mail. Otherwise, the company runs the risk of their mail becoming invisible and that the subscriber requests to unsubscribe from their database.
- The field "subject" must not be too long or complex, although it allows for the inclusion of special characters such as emojis.

There is no exact formula to predict the success of a newsletter. Each campaign is different and is closely linked to the audience it addresses. In any case, campaigns may be optimised using A/B tests.

These tests are random experiments with two differential variables. Two different versions of the newsletter are sent to two subgroups of the database just changing a single element (the field "subject", layout or organisation of content, the colour of the download button, the size of the main image, etc.). This process allows for an empirical test that reveals which version of the message works better and leads to higher ratios of opening, clicks, conversions, etc., so that optimisation of future versions of the newsletter becomes possible.

5.6 Basic Metrics to Assess the Efficiency of E-mail Marketing

There are three elements to assess e-mail marketing: the user database, sending the newsletter and conversion metrics.

In metrics related to databases, estimating the index of increase in subscribers is easy or, if that were the case, the rate of unsubscribers. Identifying the cause for any increase or decrease in subscribers is of essence. As they happen in a timeline, the cause can be traced back to a particular content item, so that the organisation knows what content works better.

Regarding sending newsletters, there are four indicators to measure the success of e-mail marketing:

- Sent index: Percentage of deliveries to the recipient (i.e. where there was no mistake in sending).
- Opening rate: Percentage of e-mails that have been opened by the recipient. Some applications provide information about what recipients have received and opened the e-mail, as well as reception and opening time.

- Clicks on links: It is possible to determine what are the links that have been clicked and establish a popularity ranking depending on the number of clicks.
- Unsubscribers per batch sent: Number of people who have cancelled their subscription to a newsletter after they have received a particular issue.

Finally, a conversion rate can be established between the sending of the newsletter and the accomplishment of a particular goal, for example downloading a mobile application, using a discount voucher, registration on another website and sending additional information by the user.

5.7 Applications and Resources for an E-mail Marketing Campaign

The supply of existing applications around e-mail marketing is wide, with many applications for different operating systems, such as Windows (G-Lock EasyMail7, for example) or OSX (Direct Mail, for example), although market trends seem to focus on the development of Web applications that help manage this kind of online marketing. The advantages of Web services lie in their mobility and the possibility of accessing them from different devices.

Thus, we can have a wide array of Web services to implement an e-mail marketing campaign successfully. The features vary a little across platforms, although there are a series of functionalities that must be present for a proper professional development of such marketing:

- Contact and list management, with the possibility to import and export, and to create segmented contact lists.
- Newsletter design, usually HTML based. Many of the tools incorporate visual editors to make the design aspects of the newsletter as easy as possible, to include multimedia elements and to distribute the elements in the newsletter.
- Sending campaigns and the possibility of scheduling and automating the process.
- Statistical analysis of output.

Some tools enable the integration of the newsletter with other services and platforms regularly used by organisations. Some examples are the incorporation of a widget on the company's blog, the integration of social media or tools such as Google Analytics.

The most widely used newsletter editors are Teenvío, Dopple, MailChimp, Benchmark, MailRelay, Campaign Monitor, MPZ Mail or SendinBlue.

These kinds of services tend to offer freemium business models, offering a limited free version that allows to test the service with some restrictions; they are usually linked to the number of subscribers and/or monthly dispatch of e-mails (see Table 1). Choosing one platform or the other depends on the needs of the company and the possibilities they offer on their free or payment versions.

Table 1 Comparison between the most important e-mail marketing applications

App	Business model	Free subscribers	Monthly mail delivery
Benchmark	Freemium	2000	14,000
MailChimp	Freemium	2000	12,000
MailRelay	Freemium	15,000	75,000
MPZ Mail	Freemium	2000	12,000
SendinBlue	Freemium	Unlimited	9000
TeEnvio	Freemium	1000	5000

Source Prepared by the authors

E-mail is a powerful marketing tool that is also easy to combine with other strategies and platforms, such as social media and e-commerce.

6 E-commerce

E-commerce means the development of transactions between companies and/or individuals on Internet, mainly for buying and selling products and services, using applications such as e-mail, instant messages, shopping carts or Web services, to name a few.

The progressive penetration of Internet and its possibilities have multiplied online trade relationships. This trend started in 1970 with the transfer of funds, with the arrival of the World Wide Web, and it became an opportunity for traditional companies and the beginning of businesses operating only online.

The boom of mobile devices, smartphones in particular, has opened the doors to mobile commerce (m-commerce), i.e. electronic commerce using a mobile device [23].

E-commerce is an increasingly wider and more diverse phenomenon; therefore, classifications are difficult. In any case, the most widely used classification of e-commerce is based on the nature of its transactions looking at the relationship between companies and/or between them and their final customers. In this line, we can speak of:

B2B: Business to business, these are already established companies that operate through Internet where consumers are not involved.

Car manufacturers, for example, use online purchase platforms for their suppliers to place their orders.

B2C: Business to consumer, this is the most popular and widespread model on Internet, where a company sells its products (e.g. Zalando, Amazon, AliExpress) or services (Spotify) to the consumer through the Web.

B2E: Business to employee, this is a model of e-commerce that is derived from the previous one, where only the company and its employees take part. Microsoft, for example, uses it so that its workers can order office supplies, documents and company cards. This is a type of business that generates engagement by the staff through, for example, attractive offers that push their performance, which is the reason why big companies are so keen on this model.

G2C: Government to consumer, this is a variation of B2C for the public administration. This is a model of e-commerce that is gaining progressive weight in its implementation through the e-administration. A clear example is payment of taxes through Internet.

C2C: Consumer to consumer is common in sectors where end consumers relate to each other, away from companies, for their own transactions. This is a business model that is becoming stronger with crowd sourcing economies (e.g. BlaBlaCar or Uber) or online purchases and sales portals (eBay).

6.1 E-commerce Techniques

6.1.1 Online Shop

In order to purchase or sell on Internet, an online shop is not a must, but it becomes a useful Web tool for efficient e-commerce [24]. Some of the most common sections in any online shops are as follows:

- Properly structured product catalogue with sections, images displaying the product alone, as well as accurate and honest product descriptions.
- Shopping cart, always visible so that the customer can check, with just one click, the products added, shipping costs, taxes and discounts (if applicable) and the total purchase price.
- An internal search engine, which will be more important as the product catalogue and sections develop.
- Explanation of payment methods allowed and contact/customer support area. As there is no physical contact, companies must offer communication pathways for their online shops so that users can share their concerns and increase their trust in the company.
- The availability of several payment options is recommended. Some of the most common payment methods are credit card, which needs a safe payment gateway, bank transfer or other e-services of great penetration and reliability such as PayPal. Payment against delivery, which was a star method for distance sales, can also be used on some shops, although it is becoming outdated (Fig. 6).

Apart from the quality of the product that the description and the picture on the online shop match the product, companies must be very careful with the shipping options, especially in the case of physical products. In online sales, transportation and the delivery of the product become the only phases where the customer has physical contact with the company, and therefore, a bad experience with the courier or deficient packaging can ruin the purchase experience.

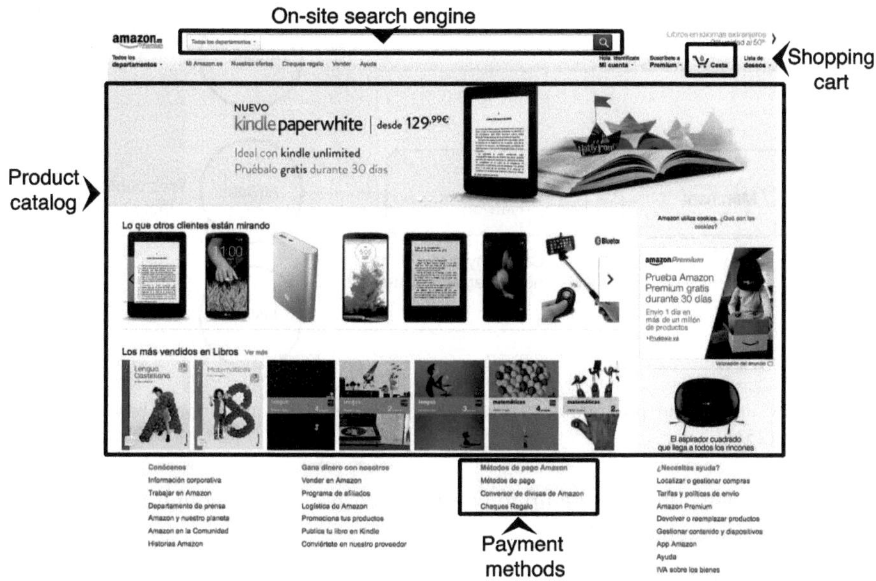

Fig. 6 Main sections of an online shop. *Source* Prepared by the authors using a screenshot of Amazon.com

Companies must take care of these aspects to the last detail, as they are part of the image the consumer will have of their brand. Guarantee and delivery dates by the courier must be considered, as they are key e-commerce issues.

6.1.2 Development of an Affiliation Programme for the Online Shop (Affiliation Marketing)

Affiliation marketing is an online marking tool widely used for e-commerce. In essence, it means taking the commission business model to an online environment. Amazon was a pioneer in this type of marketing when, in 1996, it allowed other websites to sell their books in exchange for a percentage of the unit sold.

Affiliate networks appear in order to regulate trade relations between merchants —the real product sellers—and affiliates—those who publish the ad of a product online. Such networks include Zanox, TradeDoubler or Commission Junction, and they act as mediators between advertisers and affiliates, while they provide them with tools to follow up sales and the proper development of their relationship. The affiliate network is supported by a commission that the merchant pays the affiliate (Fig. 7).

For the merchant, the advantages are obvious—they only pay if the target is accomplished and it can reach its potential clients through the hundreds of Webs promoting them. However, some affiliate networks demand a monthly fee—and

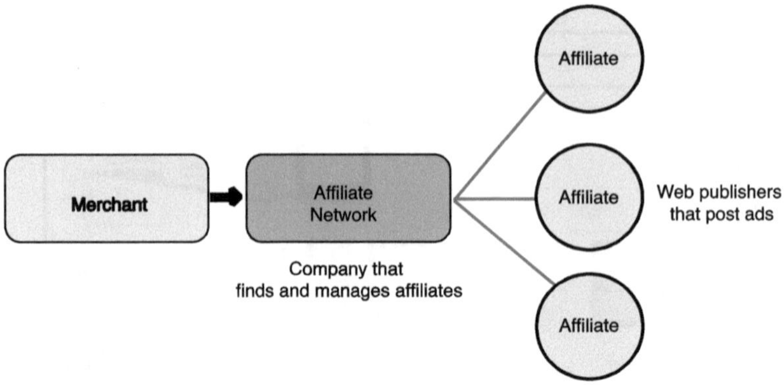

Fig. 7 How an affiliate programme works. *Source* Prepared by the authors based on Quirk eMarketing [25]

even an entry fee—that can be up to 600 €, a threshold high enough for small- and medium-sized merchants [26].

6.1.3 Retargeting or Remarketing

Retargeting is an online marketing technique to turn a user into a customer who, despite having shown some interest on the products or services of a website, did not manage to finish the purchase or action required.

Google labels this technique "remarketing" and uses it through their display network.

Retargeting works as follows: The user visits a product X on an online shop. Without finishing the purchase, they leave the website and continue browsing the Web. When they access another site, the user will find adverts on product X, adverts that will "follow them" during browsing and—in case they click—will bring the user back to the initial online shop (A) (Fig. 8).

Retargeting is only used for consumers who have shown any interest in a product before, and therefore, it is a quality impact that often ends up in a purchase.

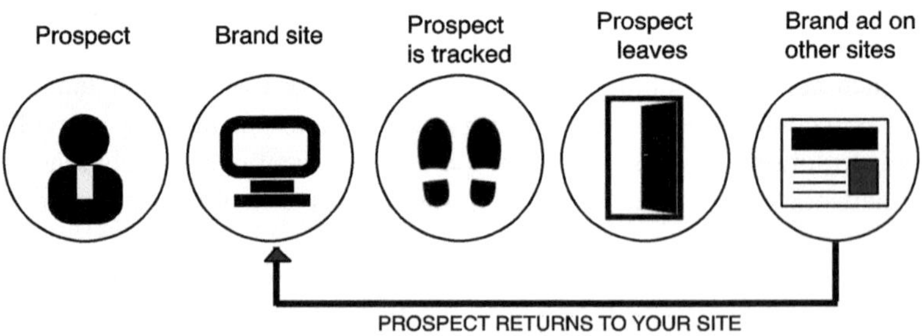

Fig. 8 How retargeting or remarketing works. *Source* Prepared by the authors based on Hussain [27]

Retargeting is part of "behavioural marketing" and is supposed to yield high profitability for e-commerce.

6.2 Business Models to Estimate Payment Per Page in Advertising

Many of the online marketing strategies are based on digital advertising; therefore, it is essential to know the different formulae to hire and estimate the cost of an online campaign. Some of the online advertising models available include the following:

- CPM (cost per 1000). This is related to the number of times an ad is shown on the screen, also known as impressions. The CPM indicates the cost of 1000 online impressions of the ad. This system is basically used for branding campaigns, and the process of brand equity is the most economic option of all.
- CPC (cost per click). This is related to the interactions of the user with the ad through clicks. They are used to divert traffic to a website, paying only when the user clicks on the ad and is redirected to a Web (therefore the name PPC, Pay Per Click). CPC does not guarantee sales, but it ensures traffic and is less volatile than CPM.
- CPL (cost per lead) refers to a contract based on quality contacts without implying direct sales. In particular, CPL is the price that is paid for each user who completes the objective or lead. Such leads vary depending on the marketing goals of the company; a lead can be to fill-in a Web form, becoming a follower of the company on social media or to disseminate content on the site.
- CPA (cost per acquisition) is hiring ads per sales; that is, payment is done for each action that has generated a customer. In the mobile environment, this is also referred to as CPI (cost per install) and indicates the applications installed after interacting with the ad. In this case, the installation of an APP, even if it is free, becomes a purchase (Fig. 9).

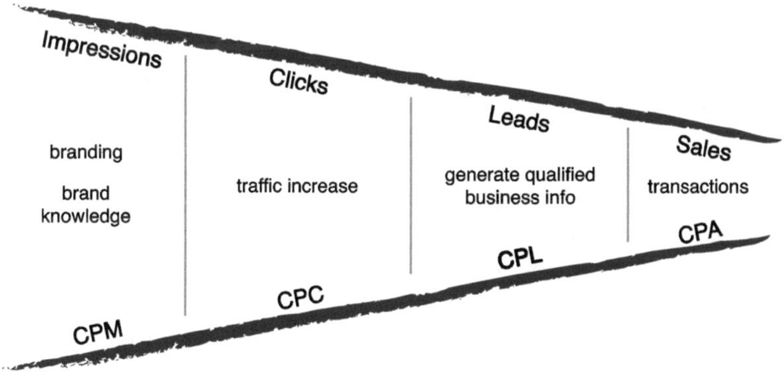

Fig. 9 Models of online ad purchasing. *Source* Antevinio [28]

Of the previous models, only CPA and CPI ensure a transaction between the customer and the company, so that they are formulae that require more economic investment.

6.3 E-commerce Glossary

Understanding the language of e-commerce is not simple. The use of common words with a different meaning can sometimes be confusing.

Some of the most usual terms are as follows:

- A/B test: through this technique, organisations show two different versions of the same content to understand which one is more widely accepted. A/B tests are common in e-commerce and online marketing and they must be done with just one change every time the test is performed.
- Shopping cart abandonment: this is the moment a potential customer gives up before finalising the purchasing process on an online shop.
- API: a set of operations and instructions released by software to interact with it and access higher quantity of data and options.
- Backoffice: administration of the backoffice of an online shop. This is mainly geared towards catalogue and stock management to optimise the browsing process and purchase of a product on the said shop.
- Call to action: this is an initiative to create interest among users and encourage them to participate or react before a particular stimulus.
- CAPTCHA: Turing test inserted on a website, generally on a form, to check whether the data are being fed by a person or a machine. It is useful to avoid spam.
- Shopping cart: a key tool of an online shop showing the products a user has selected for their purchase, their price and taxes, as well as the final cost for the user.
- Cash flow: same as for traditional businesses, e-commerce needs to look at its cash flow or the difference between receipts and payments of a company in a particular period.
- CMS: Content management system that allows, in a simple way, to organise, treat and publish on a website. This is also used for the online product catalogue and, generally, for any content on the online shop.
- Cookies: website information stored in the browser enabling better understanding of the user through their browsing (habits, interests, etc.). This is key for retargeting strategies.
- Checkout: guided process of finalising a purchase that converts the content of the shopping cart into a real sale.
- Display network: it is Google's affiliate network with over two million websites available.

- Dropshipping: type of retail sales in which the retailer does not have the actual stock of the product and issues the purchase order to the supplier once the shopping process is over. This is especially relevant for an online shop environment because it saves stocking costs.
- Eye tracking: technique and instrument to eye track the areas of the screen users pay special attention to, as well as their reading line.
- Lead or Conversion: each of the concrete goals set by the company. Usually, a lead is equivalent to a sale, but this is not necessarily so; they can be linked to a database of subscribers and prescribers.
- Payment gateway: it triggers payment processing.
- Payment processing: it allows for payment management.

7 Web Positioning

Accessing information on Internet is done mainly through the World Wide Web, a vast digital arena that has experienced constant expansion since it was created in 1991 by Tim Berners-Lee. The current volume of websites is around 850 million (July 2015).

When Internet started, the number of websites was low enough to access all of them through their domain. As Internet expanded, it became clear that there is a need to create a system that allows for Web searches and enables access.

Until the end of the twentieth century, classification systems with embedded categories, known as directories, offered good results in an expanding community of Web users. The most widely known directory was Yahoo, still accessible on the website https://business.yahoo.com/.

However, a more intuitive and simple way to access the increasing volume of sites and Web pages was needed. That was when browsers first appeared.

A browser or search engine is a computer-based system that indexes websites using some dedicated software. A search engine offers a list of the results depending on the search terms and connectors used by the user to access the desired content.

This simple process is currently the most important pathway to access online content, while it becomes the marketing base for browsers: SEO and SEM.

SEO stands for search engine optimisation, and it refers to a set of techniques applied on a website—structure, code, content and links—to improve positioning in the organic results of a concrete browser [29].

SEM stands for search engine marketing, and this refers to a publicity system of a browser offering users ads that are related to their search terms. Unlike SEO, SEM offers induced results (payment), even if they are presented to the Web surfer as the best result of their search (with the same look as for natural results) (Table 2).

Both types of positioning are opposing, even if they are both based on two key aspects: search engines and keywords (Fig. 10).

Table 2 Differences between SEO and SEM

SEO	SEM
Mid- and long-term results	Immediate results
Results are sustained for longer	At the end of the campaign, the results are erased
Organic results	Paid results
Results are always shown at the centre of the screen	Results appear mainly on the screen top or sides

Source Prepared by the authors

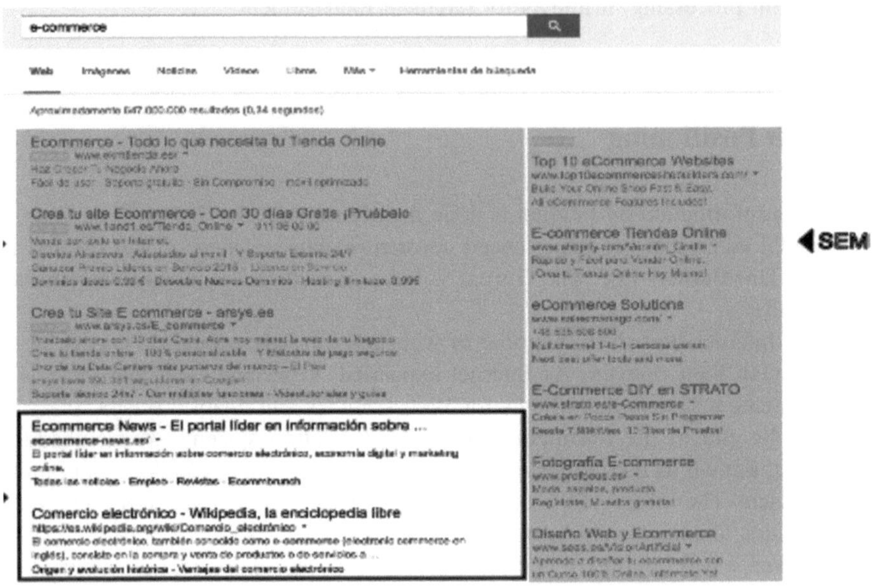

Fig. 10 How search engine optimisation and search engine marketing results are shown. *Source* Prepared by the authors

7.1 Search Engines

Each search engine uses tracing software for websites for their indexing. The software used is known as Internet bot, Web crawler or Web spider, and periodically tracks the Web looking for new content or modifications on sites.

The search engine processes Web tracking, while it uses its own algorithm to classify websites and offer its results, as well as their order, depending on the search terms used.

Therefore, both in order to improve the organic result (SEO) and to improve the efficiency of advertising campaigns (SEM), selecting a specific search engine is of essence.

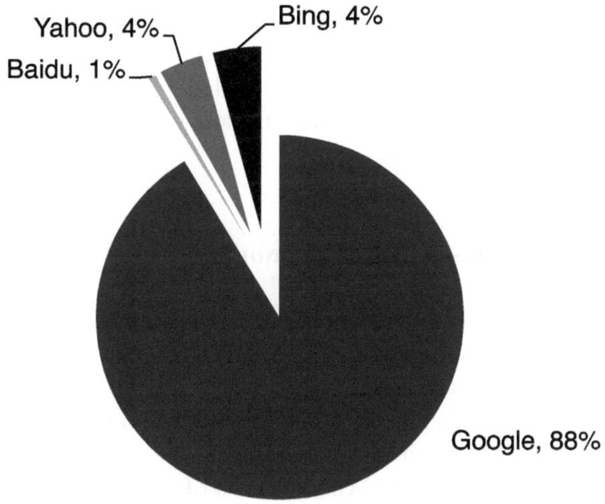

Fig. 11 Penetration share of Web browsers worldwide. *Source* Prepared by the authors using Statista [30] data for July 2015

Despite the fact that there are several search engines available (Ask, Bing, Yahoo!, AOL, Baidu, Yandex, etc.), Google ranks as the search engine with the deepest worldwide penetration.

In fact, Google is the search engine with the highest market share in most countries in the world. In countries such as Japan, Russia, China and North Korea, it ranks second.

In order to optimise resources for positioning, organisations need to adapt to the search engine's algorithm most widely used when speaking about positioning in browsers, usually meaning SEO and SEM action in Google (Fig. 11).

Google started in 1998, and in a few years, it became a world reference, thanks to an innovation they introduced in the result system: PageRank. This algorithm gives marks to each page depending on the quality of their incoming links; thus, the results are more closely adjusted to Web searches.

In recent years, Google has introduced changes in its algorithm in order to avoid anomalous positioning and bad practices in its results. Thus, Panda (released in 2011) and Penguin (released in 2014) have refined their positioning criteria.

7.2 Keywords

Keywords are the terms or phases that users use to search something on Internet. Likewise, keywords are search criteria matching a specific website.

When trying to positioning a site using some search criteria, several words are used for each keyword, avoiding generic terms at all times. These words will be increased and/or modified with time.

The extent of use of these keywords on a particular website is also important, as it is checked against the total number of words on the Web page. The percentage obtained is known as keyword density, and it must be over 2 % and below 5 %.

Likewise, in many sectors, several websites want to position themselves with the same keywords. The degree of competition to position a website over another one using the same keywords is called keyword competence. The SEM section is particularly relevant, as there is more competition and this will increase the prices to get those keywords.

7.2.1 Tools to Select Keywords

In order to select the keywords, a manual list is prepared with the marketing and site goals to be optimised. This manual list can be filtered and improved using some useful Web services.

Google Trends is a Google tool showing quantitative data regarding term searchers by their users.

The evolution of Web searches of one or several terms can be observed, thanks to this service by using three parameters:

- Time trend or interest through time, from 2004 until the present;
- Seasonality throughout the year, to be chosen since 2004 (the year the service started);
- Location, so that the result can be shown in global terms or per country.

Google Trends provides information regarding how certain search terms behave in the long run, and this is quite useful to discard or optimise a list of keywords before implementing them on a website.

Google AdWords is a PPC (Pay Per Click) advertisement management platform by Google and therefore the SEM of this browser. This platform includes a keyword measurement tool called Keyword Planner.

Keyword Planner offers information on the potential performance of a term list with data regarding the average monthly searches and competence of keywords for SEM campaigns. Due to this, Keyword Planner becomes an added value tool to select keywords.

Other interesting tools to plan keywords are Ubersuggest, which suggests alphabet-based replacement terms for a keyword; Soovle to suggest added value keywords for the most important search engines; or SEMRush to monitor competitors.

7.3 *How to Use SEO Techniques*

SEO positioning is improved by working on two aspects: internally or externally, depending on the control level that the Webmaster has over the website.

Internal SEO means undertaking actions to improve content, code and/or accessibility, aspects related to the website to be controlled by the Webmaster or company. Regarding this, keywords must be adapted to the website [31]. Such keywords must be included on:

- Title tag: the website title, visible on the top bar of the browser.
- Meta tags: even though they are losing importance (especially meta tag keywords), meta tag description is very useful, both for Web crawlers and for users, as the content of that tag is used as a rich snipped in the browser results.
- Links: using keywords on friendly URLs.
- Content: keywords must be within the content of the website. In order to be considered relevant, the content of each website has to be over 300 words and the position and density of keywords have to be looked into. Regarding location, in order to optimise the use of keywords, they have to be placed near the Web heading and on the title or subtitles of the text. Likewise, in order to improve positioning, the density of keywords should be over 2 % but not over 5 % so that browsers do not penalise the site.

With the latest update of Google's algorithm, content has become more important. Offering original and quality content is one of the recommendations by Google for those who want to get started in the optimisation for Web searches.

Apart from selection and use of keywords, improving Web positioning requires the development of other actions on the website's structure and code.

For example, having a sitemap helps Web crawlers index the website. A sitemap is a file with different URLs that make up the website.

Another action in this line is to indicate to Web spiders which content is relevant and which is not, so that they trace it (or not) and index it (or not). This action is done through the robots.txt file, located on the root directory of the website.

On the other hand, external SEO focuses on aspects that are less controllable for the company, for example incoming links (coming from other websites), with the goal of gaining popularity and higher quality of the links. A link is perceived as a quality link when it is created by a site or reference platform from the sector the company is in and/or that works on the same topic.

Since 2012, Google penalises malpractice related to external links, for example link farms (group of websites that all hyperlink to every other site without having any type of theme or sector link); therefore, it is important to work in order to create quality links to other websites relevant for the sector.

This technique is known as link building and can be implemented through concrete actions such as:

- Exchange of links between different websites sharing themes and content.
- Links from social networking sites such as forums or social networks, adding the link and the signature to the message or the content shared.
- Link baiting: publishing content that encourages visitors to create links from their websites to that content. The key to success lies in their viral or mimetic nature, and though it is a technique that is difficult to master, the results are outstanding.

Quality external links could also be improved by registering a website on general or theme-based directories. This practice is no longer used since Google stopped considering them in their update of the algorithm in 2013.

7.4 How to Use SEM Techniques

SEM campaigns require being familiar with the keywords that are related to the business one tries to promote. Each click of a user who is not part of the target is lost money.

In order to optimise a marketing campaign using browsers, one has to know the structure of an ad. An ad usually has a title of less than 25 characters linked to the destination site: a visible URL, which need not be the destination site URL, and two lines of text, around 35 characters per line. The content of those texts must be carefully decided as it must create interest in users, especially in the body of the message as the ad will be placed on locations where only a single line will be visible.

The most important ad platform for browsers is Google AdWords, which operates its own search engine and a network of associated websites (display network). In order to have SEM in Yahoo or Bing, the two other most widely used search engines, Bing Ads has to be used.

In order to successfully develop a SEM campaign, one has to consider the following questions:

- An ad must have the keywords of the destination site. If keywords that are different from the content of the website are bought to raise interest in users, many hits will be registered, but they will lack any value and shall only waste resources.
- Ads ready to pay more per word will be placed at a better spot. The value of keywords correlates with demand.
- Ads with more clicks are located in a better position.
- When designing a text, the most important aspect is that it matches the searches of users, in order to optimise the click ratio: concordance can be wide, of a sentence, exact or negative (Table 3).

7.5 Malpractice: Black Hat SEO

The importance of search engines as ways to access information of a website has opened up a space for a new professional profile: optimiser of search results—organic or paid—to increase traffic to websites.

Table 3 Keyword matching options

Match type	Special symbol	Example keyword	Ads may show on searches that
Broad match	None	Women's hats	Include misspellings, synonyms, related searches and other relevant variations
Broad match modifier	+keyword	+women's +hats	Contain the modified term (or close variations, but not synonyms), in any order
Phrase match	"keyword"	"women's hats"	Are a phrase and close variations of that phrase
Exact match	[keyword]	[women's hats]	Are an exact term and close variations of that exact term
Negative match	−keyword	−women	Are searches without the term

Source Google [32]

These experts master the algorithm of browsers and can make the best of the websites. This mastery allows them to know the limits of the browser and implement anomalous SEO improvement.

Black Hat SEO is a set of practices that try to illicitly improve the positioning of a website in the search results.

Such practices are against the standards of search engines. In this sense, each update of the search algorithm penalises some of these practices and finds out which websites have improved their positioning using them.

One of the oldest Black Hat SEO practices is to place keywords or links of the same colour on the website's background. The text, invisible for users, is indexed by search engines, increasing the density of keywords and the number of outgoing links. This practice is called hidden text and lost momentum when Google started penalising keyword density over 5 %.

Another type of Black Hat SEO practice are the above-mentioned link farms, a set of artificially connected websites through incoming and outgoing links. This practice tries to manipulate the popularity of these websites presenting them to the search engine as more relevant sites and, therefore, suited for better positioning.

8 To Summarise

In this context of intense digitalisation of individuals and organisations and of the relationships between them, it is often the case that Internet presence is perceived as a must. This is, however, not necessarily so.

Being in Internet demands constant updates of the different channels with interesting content for the public. Likewise, conversations on social platforms happen 24/7, so that a brand cannot just publish, and they have to follow the social conversation around their publications and even stimulate the participation of users to achieve higher engagement. All this needs resources that not all organisations have.

In this sense, the must is not to be on the Web but to listen to it and do it actively. In the current context, no company, regardless of its size, can live with its back to Internet.

If they do not have enough resources to have an active presence adapted to the digital world, at least they need to have a strategy to monitor and know what is being said about their company on the Web.

In this chapter, we have presented some of the most widely used and most effective digital marketing strategies. Such action can be scaled up and adapted to any brand and organisation that wants to have an active and effective online presence.

We have discussed some techniques to attract users and the right audiences, to create attractive and interesting content so that customers remain loyal to organisations and brands. While we have also presented some concrete strategies and techniques to measure outcomes, organisations working in the digital environment must be careful not to get lost in the complex universe of tools and platforms. They must give and receive value, and their most important asset is in their real essence and audience, regardless of the channels used for this relationship.

9 Review Questions (True or False)

1. The goals of a digital marketing plan must be SMART—specific, measurable, assignable, realistic, time-related.
2. Own Media is essentially online word of mouth: mentions, shares, reposts, reviews and recommendations.
3. A Social Media Plan has nothing to do with a marketing plan; it is a different document altogether and not at all related to marketing.
4. KPIs are useful metrics for objective measurement.
5. E-mail marketing is a mass, cheap and highly customised tool—the keys to its efficiency.
6. Retargeting is an online commission-based business strategy.
7. CPA is cheaper than CPM as a business model to estimate ads.
8. Google Trends and Google Keyword Planner are useful apps for selecting keywords.
9. Search engine ads related to a Web search are known as SEO.
10. Nobody can escape the Internet.

Answers to the Review Questions

1. True
2. False
3. False

4. True
5. True
6. False
7. False
8. True
9. False
10. True

References

1. Toffler A (1980) The third wave. New York: Bantam Books
2. Kotler P (1986) The Prosumer Movement: a new change for marketers. NA-Advances in consumer research, 13:510–513
3. Shultz C, Holbrook M (1999) Marketing and the tragedy of the commons: a synthesis, commentary, and analysis for action. J Public Policy Marke 18(2): 218–229
4. Kotler P (1967) Marketing management: analysis, planning, and control. New Jersey: Pearson Prentice Hall
5. Rust RT, Lemon K N, Zeithaml VA (2004) Return on marketing: using customer equity to focus marketing strategy. J mark 68(1):109–127
6. McCarthy EJ (1964) Basic marketing, a managerial approach. Illinois: R.D. Irwin
7. Lauterborn B (1990) New marketing litany: four Ps passé; C-Words take over. Advert Age 61 (41):26
8. American Marketing Association (2013, July) Definition of marketing. https://www.ama.org/AboutAMA/Pages/Definition-of-Marketing.aspx. Accessed 31 July 2015
9. Krishnamurthy S (2006) Contemporary research in e-marketing. Hershey: Idea Group Inc (IGI)
10. Osterwalder A., Pigneu Y (2010) Business model generation: a handbook for visionaries. New Jersey: Wiley
11. Doran GT (1981) There's a SMART way to write management's goals and objectives. Manag rev 70(11):35–36
12. Strong EK (1925) The psychology of selling and Advertising. New York: McGraw-Hill
13. Rogers (2011) The network is your customer: five strategies to thrive in a digital age. New Haven: Yale University Press
14. Lieb R, Owyang J (2012) The convergence media imperative: how brands must combine paid, owned, and earned media. Altimeter. http://es.slideshare.net/Altimeter/the-converged-media-imperative. Accessed 20 June 2015
15. Campos-Freire F (2008) Las redes sociales trastocan los modelos de los medios de comunicación tradicionales. Revista Latina de Comunicación Social 63:287–293
16. Boyd DM, Ellison NB (2007) Social network sites: definition, history, and scholarship. J Comput-Med Commun 13(1):210–230. doi:10.1111/j.1083-6101.2007.00393.x
17. Sheth JN, Sisodia RS (2015) Does marketing need reform?: fresh perspectives on the future. New York: Routledge
18. Baston R (2012) 8 profesiones y 7 organigramas para social media. http://elogia.net/blog/profesiones-organigramas-social-media/. Accessed 31 July 2015
19. Sanagustin E (2013) Marketing de contenidos. Estrategias para atraer clientes a tu empresa. Madrid: Anaya Multimedia

20. Ries E (2011) The lean startup: how today's entrepreneurs use continuous innovation to create radically successful businesses. New York: Crow Bussines
21. The Radicati Group (2015). Email statistics report, 2015- 2019. http://www.radicati.com/wp/wp-content/uploads/2015/02/Email-Statistics-Report-2015-2019-Executive-Summary.pdf Accessed 31 July 2015
22. Godin S (1999) Permission marketing: turning strangers into friends and friends into customers. New York: Shimon & Shuster
23. Laudon KC, Traver CG (2012) E-Commerce 2012 (8th revised edition). Boston: Prentice Hall
24. Korper, S., & Ellis, J. (2001). The E-Commerce Book: building the e-empire. San Diego: Academic Press
25. Quirk eMarketing (2012) Online Marketing Essentials. Lardbucket. http://2012books.lardbucket.org/pdfs/online-marketing-essentials.pdf Accessed 31 July 2015
26. Jimenez R (2009) Conocer las redes de afiliados. In E. Sanagustín (Ed.) Del 1.0 al 2.0, Claves para entender el nuevo marketing. (pp. 90-101). Bubok Publishing. https://app.box.com/shared/tgoujqjm72 Accessed 26 July 2015
27. Hussain K.(n.d.) Remarketing for Search. http://kirinhussain.com/remarketing-for-search/#prettyPhoto Accessed 31 July 2015
28. Antevenio (2015) Diferencias entre CPM, CPC, CPL, CPA y CPI (2015). http://www.antevenio.com/blog/2015/01/diferencias-entre-cpm-cpc-cpl-cpa-cpi/. Accessed 31 July 2015
29. Enge E, Spencer S, Stricchiola J, Fishkin R (2012) The Art of SEO (Edición: 2). Beijing: O'Reilly Media
30. Statista (2015). Global market share of search engines 2015 Statistics. http://www.statista.com/statistics/216573/worldwide-market-share-of-search-engines/ Accessed 27 July 2015
31. Google (n.d.) Search Engine Optimization Starter Guide. http://static.googleusercontent.com/media/www.google.com/es//webmasters/docs/search-engine-optimization-starter-guide.pdf Accessed 26 July 2015
32. Google (n.d.). Using keyword matching option. https://support.google.com/adwords/answer/2497836?hl=en Accessed 27 July 2015

Part II
Speaking About Human Resource Management

Human Resource Management: An Operational Perspective

Carolina Feliciana Machado

Abstract The main aim of this chapter is to give some ideas about the relevance of human resource management (HRM) in nowadays organizations, with a special emphasis in the operational perspective of HRM. So, and beginning with a glance about the evolution and concepts, giving relevance to the transition from personnel management to HRM, the chapter follows with the discussion of some of the challenges presented to this management and the presentation of the main policies and practice of HRM that can be developed by the different types of organizations.

1 Introduction

Taking into account the importance that people have in nowadays organizations, the main aim of this chapter is to give some concepts and the main ideas about the importance of HRM in competitive organizations. Indeed, and considering that all of us need to deal with people at work, it is off a critical relevance that we can have the necessary knowledge about the development of HRM in these organizations. At this level, one question can arise, namely "*HR, and consequently HRM, is an HR managers problem or, on the other hand, is a problem of all managers?*" Obviously that it is a problem of all managers, of all the organization. Although the existence of a HRM department, it is important to say that HR are not strictly affected to this department. On the contrary, HR are widespread by all organization in such a way that we can say that dealing with all the employees that work with them, all responsible are managers of these employees, and in other words, all of them are HR managers. This is why we say that HRM is a line responsibility as well as a staff function. A staff function considering that it is the HR manager who has competence to manage, organize, and coordinate all the employees that belong to

C.F. Machado (✉)
Department of Management, School of Economics and Management,
University of Minho, Campus Gualtar, 4710-057 Braga, Portugal
e-mail: carolina@eeg.uminho.pt

C. Machado and J.P. Davim (eds.), *MBA*,
Management and Industrial Engineering, DOI 10.1007/978-3-319-28281-7_3

the organization. In other words, what we want to say is that HRM is a management global philosophy that goes beyond the management of simple technical issues related to employees.

HR decisions are among the more difficult, but also the more relevant decisions that organizations need to take as they affect the organization future but also the employees own life. Nowadays, HR managers need to be flexible and with a great ability to adapt to a changing environment, as well as be proactive in identifying HR problems and, consequently, implementing new programs. HR activities assume a crucial role in the organization performance not only today, but also more and more in the future.

Beginning with a brief analysis of the evolution and HRM concept, giving some emphasis to the transition from personal management to HRM, this chapter looks to discuss some of the main HRM challenges in the second section, ending with a glance about the main HRM policies and practices.

2 Human Resource Management: Evolution and Concepts. From Personal Management to Human Resource Management

Although the human being has, since ever, belonging to organizations, and over him has been developed different action policies, it has been only from the middle of the XX century that human resource management, understood as a strategic approach of the organizations' people management, has assumed a greater relevance (with a special focus to the last two decades of this century). To Armstrong [1], its growth, as well as its future development, was deeply related to the environmental changes that were felt outside and within the organizations.

Specifically, with reference to the management thinking beginning, we can say that it was with the industrial development that was felt at the end of the XIX century, beginning of XX century, in the USA and continental Europe that it was felt the need of a greater sophistication and development of the organization traditional principles. Before that, the environment was characterized by the existence of a great diversity of organizations, with different dimensions, low income levels of the used machines, high waste levels, great dissatisfaction on the employee side, among others, facts that favored the emergence of a management science [2, 3]. In this period, a high number of employees are called to the organization, which leads managers to develop a set of principles, rules, measures, regulations, and procedures, compatible with a great employee monitoring [4]. A special focus has been given to the division of labor and worker specialization—principle that was, since a very early stage, defended by Adam Smith in 1776 in his book entitled "The Wealth of Nations"—conducive to a growing repetition and monotony of tasks. Having as main proponents Taylor [5], Ford, Gilbreth [6], Fayol [7], and Weber [8], of the Classical School, giving a special focus to the efficiency and effectiveness in the function development, the worker was, however, treated as a machine piece.

It was with the Human Relations School that it has been given a special emphasis to people in the work environment. It is now coming to be realized that it is only possible to the things through people, in other words, working with the people. We should highlight that even this School is, by excellence, that one which begun by emphasizing the human being in the organizations, this idea has been supported a century earlier with Robert Owen. It was Robert Owen who, in the beginnings of the XIX century, and in his quality of manager, began by understood the importance of the people who worked with him. Considering people as "vital machines", Owen followed the principle from which if there were a great concern with the workers, giving them better working conditions that will be extremely rewarding to the organization, at the same time that will allow to overcome the human poverty that was felt on that period.

This is why we can say that claiming for an ideal workplace to workers, Owen was at 100 years ahead when, in 1825, he wrote his book "A New View of Society." Indeed, it was necessary to wait 100 years to, at the end of the twenties; beginning of the thirties, Elton Mayo, through the Hawthorne studies, could conclude that the attention given to the social factors, at the workplace, had a relevant impact in the workers productivity levels. In the Human Relations period, many organizations created their personal departments, whose role consisted in "(…) to help keep the workers in line" [9, p. 14]. More than the simple work security and financial compensation, workers want, now, that management can see them as people, with their own needs that they want to be satisfied. It is highlighted here the importance of communication between the worker and supervisor, as well as the need to the establishment of a more participative work environment. At this level, organizations make use of their personal departments in order to create a desirable organizational climate.

Based in the Human Relations School Principles, added now by the contributions given by other scientific areas, namely psychology, sociology, political science, and among others, we face with the Behavior Science School.

Focusing, deeply, in the organization as an all, and not so much in the person itself, the Behavior Science looks to analyze the relation between the workplace and the worker, and in other words, how the workplace affects the worker, as well as how worker affects the workplace [4, 10, 11].

Understood the worker as a critical factor to the organization functioning [12–14], were many the academics that have developed a great number of action principles about this subject. Specifically, and following the work developed by Carnegie [15], Maslow [16], and McGregor [17]—to whom it is possible to add the contribution given by Peter Drucker [18]—[1]were various the psychologists and sociologists that had given their contribution, namely Fiedler, Vroom, Herzberg [19], McClelland [20], Hackman, in the psychologists groups, with contributions in the field of leadership, motivation, and positions design, as well as Perrow [21] and Pfeffer [22],

[1]It was Peter Drucker who, in 1954, in the book "The Practice of Management", for the first time had characterized the concept of "management by objectives".

in the sociologists group, contributing to a better understanding of power, conflict, and organizational design [2].

Notwithstanding the dynamic and deep interaction that is felt among all the resources that make up the organization, the idea that the human factor assumes a particular relevance can be seen as the mobilizing force of all the other resources, is reinforced by the philosophy subjacent to the sociotechnical approach, according to which, although important, the technic subsystem is not more than the responsible by the potential efficiency of the organization. The change of the potential efficiency in real efficiency only is possible through the social subsystem [23], which understood the individuals, their physic and psychological characteristics, as well as the social relations that are felt in the organization among the workers.

Following the importance that the human factor is assuming in the organizational context, also the associated personal department saw is functions changed, changing from a simple absenteeism and working times controller, and simple assistant in the recruitment processes, to the assumption of great responsibilities. More specifically, is also now from their responsibility to promote the workers socialization, training, and development. It is up to them to develop adequate processes of workers communication and motivation, promoting in them the necessary abilities and competences in order to face the environment turbulence. In summary, it is of their competence to develop a proactive working force, able to face the market challenges.

We pass, therefore, from a more rigid and close perspective of people management, to a more open, dynamic, and interactive perspective. In other words, we pass here from a personal management, oriented to the short-term, pluralist and obedient, bureaucratic mechanistic and with a strong incidence in costs minimization, to a human resource management oriented to the long-term, proactive, and strategic, promoting of a high commitment and autocontrol, unitarian, and organic flexible, seeking to usage at the maximum the capacity potential of its workers, in which it is possible and should invest [24].

The high competitiveness levels of the markets, the learning given by the Japanese management system as well as the high-performance level of organizations that gave a special emphasis to human resource management, the fall of the unions intervention, and among others constitute, according to Beaumont [24], some of the motivations that have led to the further strengthening of human resource management.

However, and although the growing relevance that has been assuming, human resource management is not yet clearly defined in the literature, not having an established knowledge body, with well-define frontiers [25, 26]. According to Popper and Guest [27] (referred by Brewster [25, p. 56]), the fluidity present in the human resource management concept "(…) is both instructive, in terms of indicating its potential power as an explanatory theory; and frustrating, in that it becomes impossible to test a theory that can subsume such a range of, often contradictory, propositions."

From the exposed results, it arises the possibility of made a distinction between two different models: a *hard* model, arising from the work developed by Fombrun,

Tichy, and Devanna in 1984 and a *soft* model, proposed (at approximately the same period) by Beer and colleagues from Harvard School [12, 24, 25, 28]. Showing a special focus in the strategy—human resource management systems should be adjusted to the strategy issues (Galbraith and Nathanson 1978, referred by Hendry [12])—the *hard* approach differs with the emphasis put in people—with their needs and personal development potential—arising from Harvard School (*soft* model).

More exactly, in the *hard* version, people are understood as simple resources, as well as many others that belong to the organization, to whom is focused a minimum cost policy. Indeed, should to be managed as any other resource of the organization, they need to be obtained at the lowest cost and used at the most possible efficiency way, in order to explore their maximum potential.

According to this approach, the key elements of the human resource policies are selection, performance appraisal, development (including training), and compensation, which allow to channeling the behavior to obtain specific objectives of development.

In summary, being very similar with the "human capital" theory [12], this approach focuses in the word "resources" that is present in the expression "human resource management" [25]. This way, and because this expression is used in detriment of the word "workers"—while people that have their own needs to be satisfied—this author has the opinion that technics such as outsourcing, franchising, and subcontract can, in some circumstances, be perfectly applicable to this approach.

On the contrary, human resource management *soft* approach focuses, essentially, in the human side of the expression "human resource management," having as source of inspiration the School of Human Relations/Human Resources. Understood the workers as a resource that must be distinguished from the others, we observe that although to some organizations they assume a higher cost, they are, however, seen as the only resource that is able to create value based on the other resources. Indeed, it is the resource whose creativity, commitment, and abilities/capabilities can originate real competitive advantages [25].

Commitment is, indeed, one of the key words that, parallel to motivation, communication, participation, and involvement, are subjacent to this approach. Assuming a special emphasis, this commitment is obtained through mutuality. Beginning from this point, Walton (1985, p. 64, referred by Hendry [12, p. 9]) has the opinion that "The new HRM model is composed of policies that promote mutuality—mutual goals, mutual respect, mutual rewards, mutual responsibility. The theory is that policies of mutuality will elicit commitment, which, in turn yield both better economic performance and greater human development."

It results that in opposition to a total interest divergence, workers direct their goals to the organization goals, assuming, freely, as theirs the goals of the organization.

In line of these ideas, workers, understood as one of the most precious resources of the organization, require most attention from the human resource function, namely a more careful selection process, more support and development, adequate compensations, as well as a correct integration process in the organization. In addition, and as a result of the importance assumed by this factor, more and more is possible to verify that human resource management, contrary to be assigned only to the specialists, is from the responsibility of all managers, whose should promote the involvement, participation, and commitment of all their workers.

Gradually, they become to be required different abilities/capacities to the different stakeholders in the organization, whose can contribute to the enter/maintenance, as well as to the exclusion from specific markets, depending, respectively, to its existence or inexistence. We can observe that contrary to be understood as a strategy follower (*hard* approach), human resource management, on the *soft* approach, is related in a more symbiotic way with the organizational strategy [25].

However, the reality where we live is provided of high levels of uncertainty and continuous change, being multiple the challenges that progressively are presented to the organizations and, within them, also to human resource management. Resulting from these challenges, human resource management cannot be guided only to the strategy/business not even to the people. At this level, Armstrong [1] has the opinion that during the last years, human resource management has been facing the challenge to direct both to the business and to the people. In other words, strategies that need to be developed at the human resources level should be in accordance with the organization general context and its goals. However, it is also important to consider that workers, having their own needs and expectations, as well as higher demand levels—both in individual and collective terms—cannot be treated as any one of the other resources.

It results that HRM action field is in a constant change. At this level, Martell and Carroll [29], Budhwar [30], Mabey, Salaman, and Storey [31], and Khatri and Budhwar [32] have the opinion that based progressively in a deeper strategic action, human resource functions, such as recruitment/selection, training/development, compensations, and performance appraisal, are in a constant interaction not only with the organizational strategy, but also with each other. We face, this way, and gradually, with the concept of integration (internal and external), as well as, and reinforcing what we have focused before, with the concept of responsibility attribution.

Highlighting this problematic Schuler [33], Brewster and Larsen [34, 35], Brewster [25], Budhwar [30], and Mabey, Salaman, and Storey [31] consider that in what concerns HRM, one of the issues that during the last years have been a focus of discussion, refers to the HRM integration in the organization business and strategy, by one side, and the attribution of responsibility to line managers (in substitution of personal specialists), at the practices/policies level that need to be developed in the HRM area, on the other hand.

Adding to these key questions, other critical elements, such as the issue that there are the workers that belong to the organization that "make the difference," considering that there are their abilities and commitment that allow to distinguish the best successful organizations from the others; and also that manage HR assume a real strategic relevance [31]; contribute to the idea that HR, and consequently, HRM in organizations can be understood as a sustainable source of competitive advantage [31, 32]. This idea is, furthermore, shared by Goshal [36] when he said that the real scarce resources are people, ideas, entrepreneurship, and knowledge.

Summary, it results that, in general, HRM expression can be understood as synonymous of strategic HRM, concept that, according to Boxall (referred by Budhwar [30]), highlight the nature, progressively proactive of the HR function, its potential relevance to the well succeed performance of the organizations, as well as the ability to change the HRM function from a reactive, prescriptive and administrative perspective, to a proactive, descriptive, and executive perspective.

3 Human Resource Management: New Challenges

From this exposition, it is possible to deduce that HRM is in a constant interaction with the organizational strategy. Reporting to the research developed by Storey and Freedman, Brewster [25] has the opinion that both in the research and in the management area is evident a growing attention in the relation that exists between the organization strategy and HRM. Reinforcing this idea, this author adds that as organizations are oriented to knowledge, service, or new technologies, more the human behavior become a competitive factor, deeply related with the organization strategic direction/orientation.

This is why we say that strategy, structure, and HR function need to be interrelated in order to obtain the organizational effectivity. Indeed, we can say that in many circumstances HR function should appear in first place [37], as far as HR quality, extremely difficult to be imitated by competitors, is seen as a critical factor in the competitive context in which the organization is integrated.

Based in a set of policies, practices, and systems that influence employees' behavior, attitudes, and performance [38, 39], HRM has been a focus of special attention in the management area during the last years. According to Fisher, Schoenfeldt, and Shaw [39, p. 5] "This increased attention comes from the realization that an organization's employees enable an organization to achieve its goals, and the management of these human resources is critical to an organization's success."

Understood as a critical factor, when effectively implemented, HRM practices are related with the organization performance, contributing to the employees and clients satisfaction, the innovation, productivity, as well as the promotion of a favorable image of the organization in the markets where is integrated [38]. According to these authors, although important, the HRM potential role, at the organizational performance level, has been recognized only during these last years.

Conscious of this problematic relevance, Mueller [40] and Huselid [41] have developed some research in order to study the relationship between the adopted HR strategies/policies and the organization performance. However, and although the constant argumentation about the existence of a direct relation between the effectivity of HRM practices developed by the organization and its performance, the results obtained by these authors do not allow us to conclude this. Showing a less optimistic position from that of Huselid—to whom that relation is automatic—Frank Mueller [40] has the opinion that effective practices of resources do not translate, necessarily, in the organization performance growth. Completing his mind, Mueller [40, p. 164] considers that in order that HR can be seen as "strategic assets," "(…) what is truly valuable is the 'social architecture' that results from ongoing skill formation activities, incidental or informal learning, forms of spontaneous cooperation, the tacit knowledge that accumulates as the—often unplanned—side-effect of intentional corporate behavior."

Without questioning the divergent opinions of these two authors in what concerns these issues, other researchers such as Newton and Findlay [42] and Storey and Sisson [43], although highlighting the difficulties in guarantee, immediately, the existence of a direct relation, have, however, demonstrated, in a well succeed way, the growing potential contribution of HRM (from a strategic point of view) to the organizational performance [14].

During this debate, we can observe that there are many the challenges that progressively are being posed to HRM, namely in what concerns the emphasis posed in teamwork, flexibility, the employees involvement, as well as in the organizational commitment. About this last one, Guest [27, 44] has the opinion that the achievement of high levels of innovation and the development of strategies based in quality, in order to be well succeed, require a workforce that needs to be committed in the organization.

Progressively, the dynamic that is felt in the environment put new exigencies to the management and, within it, to the HRM, conducting that parallel to the traditional needs—more oriented to the working physical conditions, also necessary to be obtained—HRM systems should also answer to the new needs related to teamwork, flexibility, participation, and commitment both at the employee and at the organization, as an all, level [24].

The flexible organization development assumes, nowadays, as a critical factor to organizations, demanding from management new competences and abilities. In order to face a more competitive market, it is not enough to invest in new technologies characterized by a greater flexibility. Beyond these, also the own organization and its HR should be flexible and, in other words, susceptible to adapt quickly and easily to the new market exigencies. According to Kovács [45, pp. 30–31], "Management methods and more flexible organizational forms spread more and more as they seem to be most appropriate not only to face market and technology changes, but also to satisfy HR with higher school levels and exigent in terms of their work content, responsibility and participation."

By the relevance that it is subjacent, flexibility is understood as a multidimensional concept, "demanding agility and versatility; associated with change, innovation, and novelty; coupled with robustness and resilience, implying stability, sustainable advantage, and capabilities that may evolve over time" [46, p. 198].

By opposition to a type of functional organization, where functions and tasks are separate, flexible organization supposes the existence of flexible HR, in other words, qualified HR, responsible, able to work in teams and according a polyvalence system. At the polyvalence level, and in order to better obtain it, the development of training programs assumes a special highlight.

Pioneer in the study of flexibility at the HR level, Atkinson [47, 48] has developed the "flexible organization" model, in which highlight the difference between functional flexibility, clearly expressed in the organization employees hard kernel—permanent employees, with full-time contracts of employment—and numerical flexibility that is present in the employees peripheral group, characterized by the existence of part-time, subcontracted, and temporary contracts of employment.

Doing a comparative study among different European countries, Brewster, Hegewisch, and Mayne [49] and Brewster, Mayne, and Tregaskis [50] distinguish between contractual flexibility, giving to telework a special focus, and flexible working time. Albeit to a lesser extent, these authors also focus their attention in the functional flexibility, which withdraw from the functions rigid description, characteristic of the classical management model.

Following the Atkinson's [47, 48] "flexible organization" model, Kovács [45] has the opinion that more than a technical phenomenon, flexibility consists in a technic organizational phenomenon, considering that it refers not only to the flexible equipment, but also "(…) to the ability for a rapid adaptation of employees, groups, units and the organization management as an all, to the new exigencies and opportunities," ability that can be obtained "(…) by the vertical and horizontal integration of the tasks" [45, p. 33]. Kovács [45] also appeals to the professional flexibility, which referring "(…) to the employees ability to develop different tasks with autonomy" [45, p. 33], and has subjacent the existence of a polyvalent qualification, as well as the creation of new professional profiles.

Following the organizational flexibility, we can observe a progressive vertical and horizontal integration of tasks/functions—by opposition to a traditional specialization—which are now developed not by individual employees strictly related to a certain workplace, but by teams that have under its responsibility a diversity of functions. In other words, in a flexibility context, teamwork assumes as the most adequate as long as provided its members with a polyvalent training, followed by an effective cooperation, communication, and participation; it is possible for them to identify and act quickly in the presence of possible malfunctions.

Assuming the own team a higher autonomy and the responsibility by the allocation of the respective tasks, this may bring a high level of motivation [39] and commitment [51].

Fisher, Schoenfeldt, and Shaw [39] have the opinion that team participation tends to motivate employees. Asking to employees to use their creative abilities, making them responsible by their performance, promote the enrichment of functions.

Edwards and Wright [51], conscious that commitment can be obtained by different ways—Guest [44], for example, has the opinion that HRM is widely understood as promotor of the employees commitment in organization—consider that work teams are a way, particularly developed, to promote the employees commitment. Indeed, considering that commitment is based in an auto regulated behavior—and not controlled by sanctions or external pressures to the person [52] —it is possible to observe that teamwork, based on an autocontrolled system, constitutes an excellent support to a commitment promotion.

Contributing to higher commitment levels, employees' participation, at their responsibilities level, also assumes a special focus reason why should they be one of the priorities to implement. Effectively, only with a participative management is possible to overcome the high number of resistances that are, frequently, seen in the different work situations. Felling as a part of the decision-making processes, employees show a high level of motivation, involvement, and commitment when they develop their tasks. It results that in the sequence of this participation, arising from highly oriented people management practices, all the change is understood as a change implemented **by** the people and **of** the people.

Even more, and resultant from the evolution and the dynamic that progressively is emerging in the environment, HRM has to deal with a number of challenges. Indeed, issues such as diversity management [53–56], equal opportunities at the employment level [57–59], gender problematic [60–63], sexual harassment at the workplace [64], ethics in HRM [65], international HRM [25, 66–71], and among others are only some of the multiple challenges that contribute to the HRM dynamic. More than simple themes that can be added to the Behavior Sciences, each of them represents a new way of see and understand the work, the worker, and the own world [72].

In order to deal, and consequently answer, to all of these challenges, organization should promote the existence of an open culture, interactive, where we can observe a continuous apprenticeship of all of their members.

We face here with the concept of **learning organization**, according to which the employees develop, continuously, their abilities in order to obtain the established results [38]. More exactly, and according to these authors, subjacent to learning organizations, we observe that organizations should "live" a constant apprenticeship process through environment monitoring, assimilate information and take decisions, as well as restructure in the flexibility direction, in order to compete in the environment where it is integrated. It is observed that all organizations that develop such apprenticeship abilities are promoting the creation of competitive advantages.

In view of this, results that with times development, HRM has been assuming a higher relevance at the organizational context. Effectively, from a simple administrative role, with a special focus to the hiring, control and remunerations, without any strategic sense, HRM began, progressively—and as we have seen before—to give a special focus to a great number of policies, practices, and systems that influence behaviors, attitudes, and the employees performance, namely HR planning, recruitment, selection, training and development, compensation and incentives, appraisal and performance management, labor relations, communication, leadership and motivation systems, and support and follow-up. More, understood the human being as one of the success critical factors, HR managers are progressively asked to intervene actively in the strategic decision processes, resulting from this new challenge the establishment of new exigencies in what concerns the abilities they need to have. More exactly, and according to Noe, Hollenbeck, Gerhart, and Wright [38], in these recent years, and even more in the future, HR managers, parallel to a technic professional knowledge, also need to pursue abilities related to the organization activity/business, the change process management, as well as integration abilities (namely in what concerns the integration of all the other abilities).

In all of this process, the support, follow-up, motivation, involvement, participation, and commitment of all employees are a reality widely subjacent to these new abilities.

Concluding, and taking into account Sparrow and Marchington [73, p. 313] "Commitment to a professional and proactive HR philosophy, (...), is a central feature of the new agenda."

4 Looking at the Main HRM Policies and Practices

HRM consists in the planning, organization, development, coordination, and control of a set of technics that are able to promote the efficient and effective performance of all those that work in the organization [74, 75]. More exactly, what we can say is that HRM process in an organization is comprised as shown in Fig. 1. Taking into account this figure, it is now time to analyze each of the policies/practices that are possible to be implemented at the HRM level.

4.1 Job Analysis

Whenever we want to develop an HR planning necessary to any organization, we face a first step that consists in implementing a job analysis about the functions that exist in the organization. But, what is a job analysis?

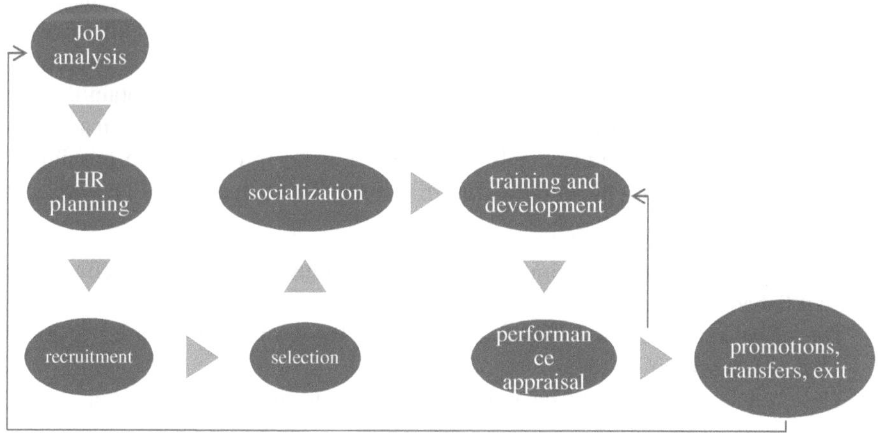

Fig. 1 HRM process in the organization

Job analysis is the act by which the organization analyzes and describes the work that its employees do. This is a way which looks to allow the organization to understand how its employees' efforts are spent by them [75–77].

So, based in the job analysis, it is possible to managers to know what are the real needs of the job, which works as a valuable tool, for instance, to the future processes of recruitment, training, and promotion. Indeed, and as it is possible to be seen below, are many the uses that can be obtained by the job analysis namely in what concerns:

- HR planning, recruitment, and selection
- Job evaluation;
- Performance standards and performance appraisal process;
- HR training and development planning
- Compensation systems;
- Employee career planning;
- New technologies design.

At this point, it is important to highlight that job analysis focuses not only the relevant issues related with the job execution (namely duties and responsibilities, tasks and activities definition, the necessary behaviors, and performance standards), but also the qualifications that employees need to pursue in order to develop their functions.

To its implementation, managers can make use of a set of methods, namely direct observation, interviews, and surveys.

4.2 HR Planning

When someone speaks about HR planning, the question that arises is to known what is its main aim, in other words, what do we want with planning? At this stage, one can say that HR planning is useful to establish the number and qualifications of the employees that the organization needs in order to reach its strategic objectives, as well as the actions that are necessary to be developed in order to obtain them in due time [75, 78, 79]. More exactly, based in planning the organization looks to anticipate which are its future needs in the HR domain in order to be possible to develop and implement efficiently its future actions. To do that, it will be necessary to fulfill three steps:

- The organizational main aims need to be previously and clearly defined and established;
- The organization needs to have a deep knowledge of itself in order to know what exists within it;
- The organization needs to have a deep knowledge of the work market.

Acting this way, the organization, and more exactly the HR department, will be able to gather the necessary conditions compatible with the assumption of a

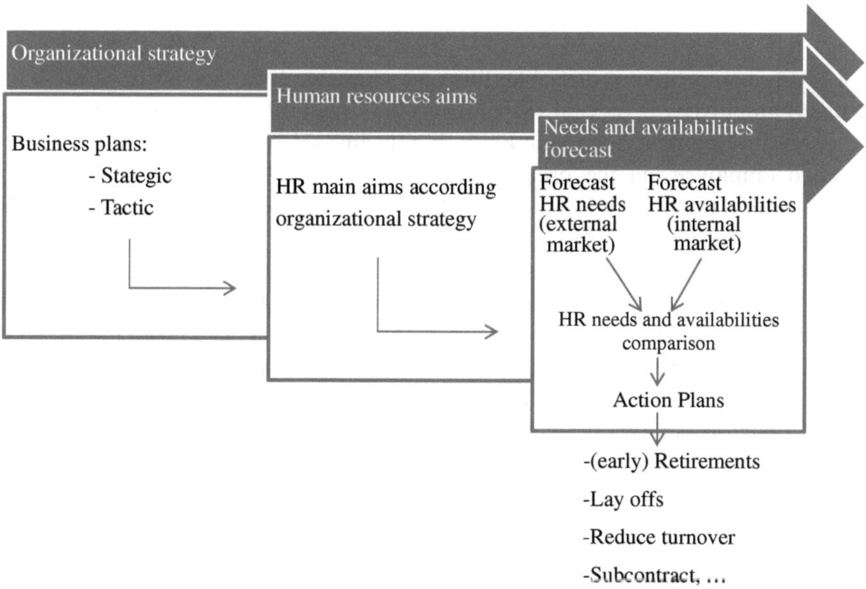

Fig. 2 Human resource planning process

proactive attitude in relation to the environment in which is integrated (and not a reactive attitude has happened in the past).

Graphically, HR planning can be presented as follows (Fig. 2).

From this analysis, it is possible to highlight that HR planning process is characterized by three distinct phases:

- HR needs forecast;
- HR demand forecast;
- Action programs development.

In order to implement these forecasts, it is possible to make use of a set of different technics, namely:

- Job analysis
- HR audit, in order to identify HR competences and abilities. This way, it will be possible to obtain a global vision of the existent talents in the organization;
- HR needs and demands statistical methods of prevision.

Once estimated the HR needs and demands, it is necessary to proceed to its comparison. Comparing HR needs (to short and long term) with internal demands, estimation will be possible to elaborate HR plans for internal fulfillment (namely training, promotions, and careers development) as well as for external fulfillment (namely recruitment, selection) [75, 78].

4.3 HR Recruitment

In a highly competitive society like this one where we live, the choice of the most qualified employees is of vital relevance to the organization development. Indeed, the good development of production is highly limited by the presence, or not, of qualified employees in the organization. In this context, recruitment assumes a relevant role, reason why it is important to understand in what consists the recruitment process.

Recruitment can be seen as a set of technics and processes that look to attract candidates potentially qualified and capable to occupy and develop different functions in the organization, among whom will be possible, in a posterior phase, to select the most adequate to the functions [76]. With recruitment, the organization disseminate and offer to the HR market the employ opportunities that need to fulfill. This is a way to open its doors and communicate with the external environment.

In what concerns the recruitment of its employees, the organization can adopt three different types:

- Internal recruitment
- External recruitment
- Both internal and external recruitment.

About internal recruitment, it can be observed when the organization looks to promote the internal mobility of its HR. This mobility can be expressed through different processes such as promotions, transfers, and transfers accompanied with promotions.

External recruitment occurs when the organization ought, or simple prefer, to go to the external work market. Based in this option, the organization looks to fulfill its functions with external candidates, people that are attracted by the different recruitment techniques. Obviously that both of them present advantages and disadvantages [80] as shown in Figs. 3 and 4.

While in internal recruitment the main techniques to promote internal mobility are job display and managers development; in what concerns the external recruitment are multiple the recruitment sources available. Particularly, it is possible to highlight the spontaneous, the recruitment agencies, other organizations (competitors or not), universities (as well as other schools), professional associations, employ services, head hunters organizations, journals and magazines advertisements, and more frequently nowadays the Internet [80].

Nevertheless, the existence of these two types of recruitment, in practice, what we can see is that organization never uses strictly the internal recruitment, nor the external recruitment. What really happens is that organizations, based in the complementarity that exist between them, prefer to develop a hybrid option as they make use to both the internal and external HR sources.

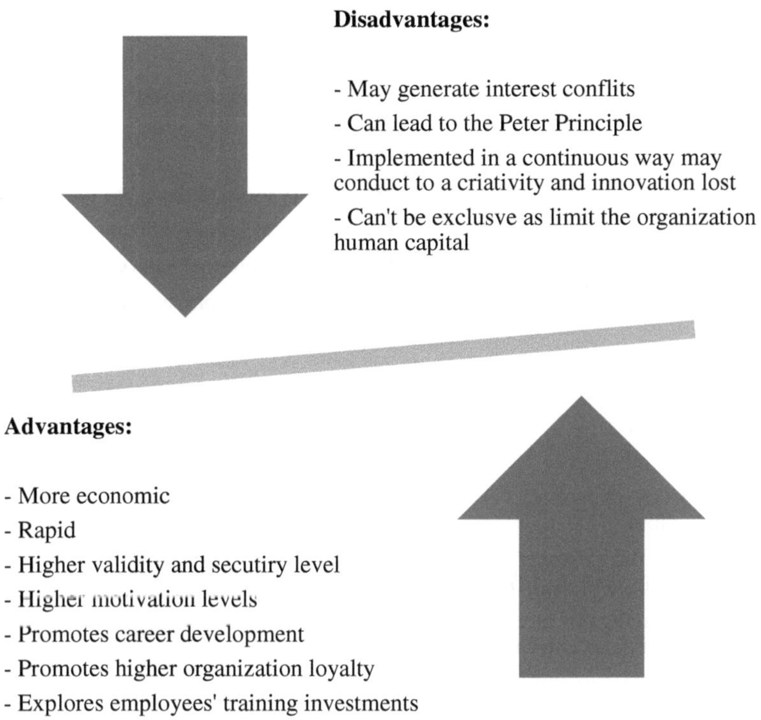

Fig. 3 Internal recruitment advantages and disadvantages

In opposition to the recruitment process, is possible to observe the HR reduction process. Very frequent in periods of crisis, needing decouple some of its employees organizations can make use of [78]:

- Tranfers: when employees are moved from one function to another. Not contributing to costs reduction, this option may, however, contribute to a reduction of imbalances within the organization.
- Dismissals: HR are permanently detached from the organization.
- Anticipated retirement: The organization gives some incentives to employees in order to leave it before the retirement period.
- Layoff: involuntary temporary dismissal. It can be observed for a few days as well as be maintained for years.
- Working hour reduction: Employees work less hours by week; divide functions with other employees; or can make use of a part-time option.
- Outplacement: when the organization looks to help its dismissed employees until they find a new job. Usually, this process occurs when, due to some crisis situations, some organizations feel obliged to dismiss some of its employees.

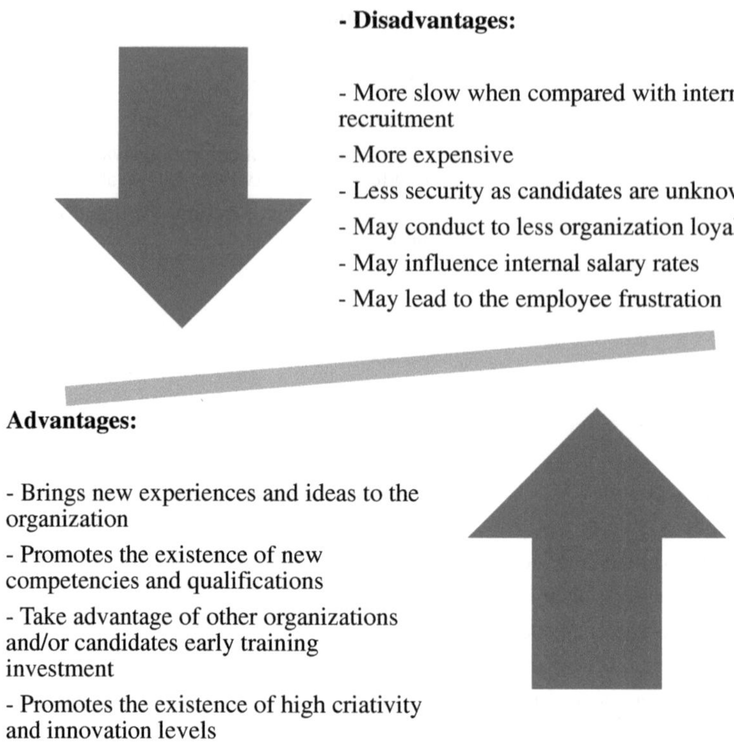

Fig. 4 External recruitment advantages and disadvantages

4.4 HR Selection

Following the recruitment process, it will be necessary to choose, within the group of candidates attracted by recruitment techniques, that or those that demonstrate higher success possibilities in what concerns the performance of the function to which the organization is selecting. Looking to choose the best and more adequate candidate to fulfill vacancies, HR selection process aims to provide the organization with the person(s) that joint the best conditions to perform the function(s), increasing, this way, the organizational performance [76]. Closely related to the organization strategy, in the selection process implementation, HR managers (selection responsible) have at their disposal a diversity of selection tools, as follows [79]:

- Application form: containing general information about the candidate, such as name, address, qualifications, last jobs, and experiences,
- References analysis: analysis of the references given by third persons/entities about the candidate,

- Psycho-technique tests: At this stage, it is possible to make use of a diversity of tests, namely intelligence tests, knowledge tests, capabilities tests, and personality tests,
- Calligraphy analysis: using biographical method,
- Interviews: Although its subjectivity, this is a selection method with a high practical application,
- Work sampling: consisting in perform a small sample of the function to be developed,
- Assessment centers: which refer to the existence of centers to assess the candidates' performance when simulating real work situations.

In practice, and in order to choose the most adequate candidates, HR responsible does not make use of only one of these possibilities. On the contrary, the most strategic option goes to the combination of some of them taking into account the content of the job to occupy.

4.5 HR Integration and Socialization

From the moment when the candidate was selected to occupy the job, it is of critical relevance to present him/her to the organization. It is necessary to proceed to his/her orientation under which the new employee has contact with the organization history, learn about its products and services, and receive information about the organization objectives, in short, and is presented to the other employees of the organization (supervisors, pairs, subordinates). Varying from a few minutes/hours, to somedays, or months, it is important that before begin effectively to perform the function it is given to the new employee the needed time to fit in the organization. Of course that the higher this period can be, more effective will be the integration of the new employee.

Jointly with the orientation also the socialization process occurs, which is seen as the process under which the new employee learn about the value system, the organization rules, norms, and the behavior standards that are required by the organization or the work group. At this moment, the employee becomes part of the organization. As a new member, he/she begins to perform the attributed tasks according to the respective function [74, 76].

4.6 HR Training and Development

Today, and even more in the future, people cannot stop learning. Resulting from the challenges and changes that we face in the present days, it is of critical relevance that everybody is able to conduct the necessary efforts in order to promote the knowledge update as a way of better answer to the exigencies originating from the respective environment. However, if, generally speaking, this is our present reality, it assumes a higher relevance in the organization context. Indeed, only with qualified employees, with the necessary knowledge and abilities, organizations are able to survive in a changing world. This is the reason why organizations have been promoting (or, at least, should promote) increasing efforts and investments in training and development programs.

But, what can we understand by training and development?

Training and development is a set of efforts developed by the organization, based in a program, which look to promote the development of the employees technical and interpersonal relationship abilities. It results that its main purpose is to improve not only the present performance, but also the employees' future performance, increasing, through the apprenticeship, their technical and of relationship potentialities and abilities [81].

Considering the importance that such a program assumes, and in order to meet the desired aims, it should be implemented the following four different steps, namely:

- Identification of training needs;
- Elaboration of the training program to answer the needs;
- Implementation and execution of the training programs;
- Evaluation of training results.

These steps can be found in the following representation: (Fig. 5)

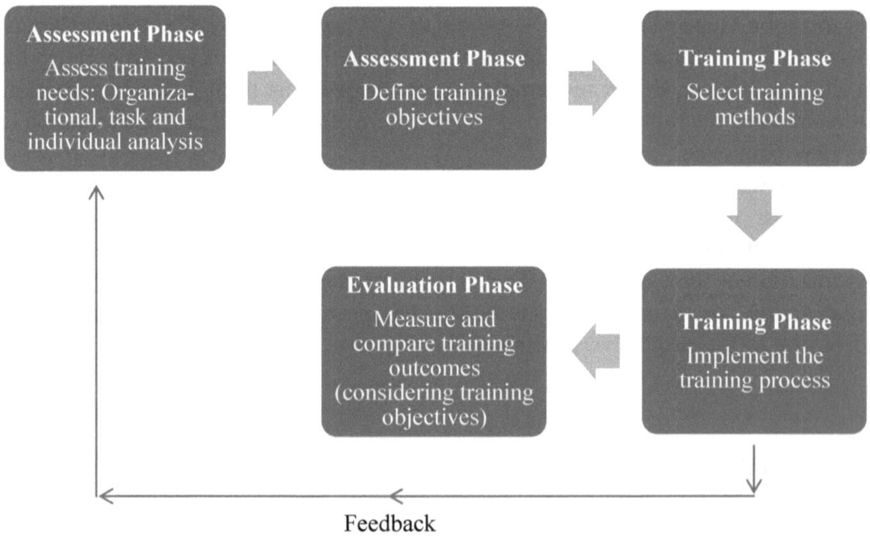

Fig. 5 Training system model feedback adapted from [82]

As it happens with any other type of investment, also here should have a monitoring and evaluation process. More exactly, in what concerns the evaluation we can say that it focuses four different levels [82]:

- Results (we want to know if the organization is better now, resulted of the training program);
- Behaviors (the purpose is to understand whether employees now have a different behavior, working better, and using the new knowledge, in their job);
- Learning (the objective is to understand whether the knowledge and abilities that employees have now are greater than before training);
- Reaction (the interest is to know the opinion that the trainees have about the training program, the trainer, the way how the training process occurs, etc.)

If, from this evaluation process, the expected results were not obtained, managers need to find the reason why it has happened in order that it cannot happen again in the future.

When we speak about personal development, in some way we have present the idea of the employee growing in the organization. Related to this growing idea, it is also possible to relate the idea of career in the organization. When an employee enters to a new organization, he does not want to stay forever in the same job. On the contrary, he establishes some aims, in terms of career, defining the most adequate course to meet these aims. Regardless of these efforts developed by employees, the organization also should define and specify its employees' career development. However, if for a long period of time was possible to speak with assurance about this issue, in other words, there exist well-defined career plans; nowadays, the reality is completely different. In fact, the greater number of employees with competencies to occupy higher hierarchical jobs, the flattening in the organization structure, and among others are factors that have been contributing to a most disturbed evolution within the organization, contrary to the linear career evolution that was observed in the past, as shown in Fig. 6.

Taking into account this new reality, and in order to avoid possible frustrations by employees, who began to understand that their career evolution is now more difficult than in the past, the organization ought to give them other evolution opportunities, such as greater responsibilities, more enriched jobs, and high participation levels.

4.7 Performance Appraisal

When an employee is performing his tasks at the organization, it looks necessary to analyze and appreciate his performance at the function as well as his performance potential. Designed by performance appraisal, this process allows the organization managers to know the employees value, excellence, qualities, and competences. This is a dynamic concept as employees have been constantly evaluated, formally or informally, in a continuous way. It is a critical tool as it allows find a lot of

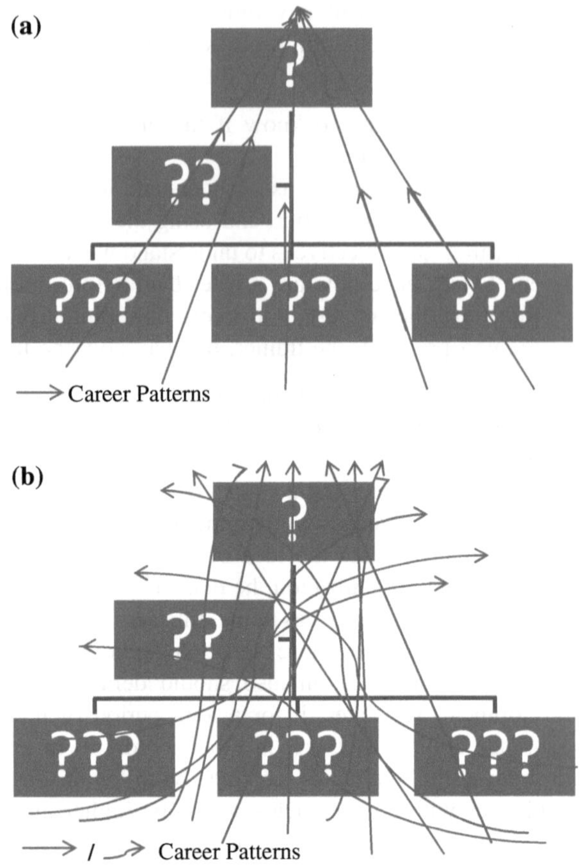

Fig. 6 **a** Traditional career flow pattern. **b** Present and future career flow pattern

problems, such as supervision, employees integration in the organization/function, motivation, and misuse of the real employees' potential. Generally speaking with a performance appraisal system, the organization will be able to reach a set of purposes like to give feedback to employees, to make more effective employees decisions, to establish objectives for training programs, and to implement more effective diagnosis of organizational problems [83–85].

It is important to highlight that this performance appraisal process is seen as having a critical organizational role only when it gives feedback that allow, all the time, to inform the employees about what they are doing and how they are doing.

Concerning to who implement this appraisal process, it is possible to say that usually this role is given to the immediate chief. However, it can also be implemented by colleagues, the own employee (autoperformance appraisal), external sources (namely clients), or even a commission specially designed to this end. More recently, the option by a 360° appraisal (through which the different persons that establish a work relationship with the employee evaluate his/her job performance) becomes more and more frequent when we speak about performance appraisal systems.

During this process, and although the assessment responsible should be neutral, the truth is that many times he is not able to assume such neutrality, reason why is common the existence of some errors in the evaluation process. Indeed, developing a very subjective evaluation contributes to the existence of systematic errors which can distort the evaluation processes. Among the most frequent errors, it is possible to highlight the errors of complacence, severity, halo, central tendency recently.

Conscious of the possibility of these errors existence, it is important that the performance appraisal is responsible to develop a set of activities before, during, and after the appraisal [86]. More specifically, it is important that before this process the responsible prepares itself to this process, namely getting training in performance appraisal methods, be aware of the main aims and strategy of the organization, and motivate the employees to the performance appraisal process speaking with them about it. During the process, this responsible should encourage employees participation, be clear and specific, be an effective listener, judge the employees performance and not his/her personality, and establish jointly with the employees the objectives to be reached in the future. Finally, and after the performance appraisal process, it is important that the supervisors/evaluators establish effective communication flows with their employees in order to give them a continuous feedback about their performance, as well as compensate these employees by their positive performance.

4.8 HR Compensation

Finally, it is important to speak a little about HR compensation.

Taking into account the work spent by the employees, organizations look to develop compensation systems in order to effectively compensate them. Along with the remuneration itself, the compensation package also includes a set of incentives and benefits. Traditionally, and very frequent nowadays as this is the most usual practice, compensations are determined based in jobs characteristics. At this level, job analysis assumes a critical role as is based on it that it is possible to identify the most critical criteria of each job and consequently establish the different salary ranges.

During the last years, however, we have been facing a diametrically opposed approach which consists in compensate employees based on their working competences and knowledge. This different approach brings a positive influence to the organizational climate as well as the employees' motivation [87]. Indeed, employees are encouraged to improve their competences and promote their development. Despite these advantages, this method presents an increased administrative complexity, originating a large increase in training and development costs.

Still related to the compensations, it is possible to add that most organizations pursue a system to compensate their employees through merit payment, which is seen as an effective system to reinforce the desired behaviors. However, it is of great importance that the design of these merit compensations is in accordance with all other organization systems. For instance, if the employee develops an individual job, compensation should be given in an individual way. However, if cooperation among the different organization employees is of critical relevance to its proper functioning, it should use team compensation plans (and not individual plans), based in team performance, as an all, and never as a simple sum of individuals. Further, a merit compensation system should be developed only after the implementation of a performance appraisal process that allows to distinguish, with objectivity and validity, the different employees' merit.

About incentives and benefits, these are understood as a set of facilities, advantages, and services that organizations offer to their employees [88]. These benefits constitute the so-called indirect compensation, which is given to all employees independently of their job and is distributed jointly with the direct compensation, that is, the job specific remuneration and result of the job evaluation or the employees' job performance.

More and more organizations are adopting the use of compensation packages which include the remuneration/salary, monetary incentives, social benefits (such as insurance, health assistance, and day care for children), status symbols, more interesting jobs, flexible working hours, and among others.

The use of "cafeteria-style" compensations is also an option that can be used by organizations. This is a kind of compensation which assumes a flexible form. The organization gives to the employee the opportunity to choose the benefit package that most interest him. According to the "cafeteria-style" system, the organization offers to its employees, for instance, two or three alternative benefit packages, where different needs are satisfied. Although this decision to implement this "cafeteria-style" system may imply some administrative difficulties, it presents, however, a high motivational power within the organization.

5 Some Final Remarks

Has we have seen before an organization is characterized by different types of resources, namely physical resources (buildings, equipment, materials,…), financial resources (money, investments, financial and capital operations,…), and human resources (people). All of them are of critical relevance to the well and normal organization functioning. However, if in what concerns the first and second type of resources, there are not many difficulties in their measurement, as we have ways to know how is their cost, the same does not happen about HR. Extremely subjective and very diversified among each other, these resources are not so easily measurable as the others. Obviously that it is also possible to determinate what its value is, namely what is the cost of its recruitment and selection, its relocation, and the training and development investment. However, there exist much more beyond these issues that are not easy to measure as they are not so evident and quantifiable to one person eyes.

In other words, what we want to say is that HR concept is an extremely broad concept, holistic, as it is present in all the organization and in a constant interaction with all its parts and components, reason why it assumes a special highlight to the organization the way how employees are evaluated and valued or, more exactly, the way how organization attract and obtain its employees and maintain and develop them within itself. HR are, by this way, one of its most critical resources, as they are directly related to the acquisition of higher and more effective levels of organizational efficiency and effectiveness. Understood as human beings that can be more or less motivated and commit with their job and their tasks, parallel to the necessary coordination, organizations ought to promote high levels of motivation and commitment.

6 Review Questions (True or False)

1. HR decisions are among the more difficult, but also the more relevant decisions that organizations need to take as they affect the organization future but also the employees own life.
2. There are many challenges that progressively are being posed to HRM, namely in what concerns the emphasis posed in teamwork, flexibility, the employees involvement, as well as in the organizational commitment.
3. Job analysis focuses particularly to the relevant issues related to the job execution, and in a marginal way the qualifications that employees need to pursue in order to develop their functions.
4. HR planning is useful to establish the number and qualifications of the employees that the organization needs in order to reach its strategic objectives, as well as the actions that are necessary to be developed in order to obtain them in due time.
5. Recruitment can be seen as a set of technics and processes that look to select candidates potentially qualified and capable to occupy and develop different functions in the organization.
6. Very objective in its nature, interviews correspond to a selection method with a high practical application.
7. Jointly with the orientation also the socialization process occurs, which is seen as the process under which the new employee learn about the value system, the organization rules, norms, and the behavior standards that are required by the organization or the work group.
8. HR training and development process is characterized by three steps, namely identification of training needs, implementation and execution of the training programs, and evaluation of training results.

9. Performance appraisal process allows the organization managers to know the employees value, excellence, qualities, and competences.
10. Merit payment is seen as an effective system to reinforce the desired behaviors.

Answers to the Review Questions

1. (True—see section 2)
2. (True—see section 3)
3. (False—see section 4.1)
4. (True—see section 4.2)
5. (False—see section 4.3)
6. (False—see section 4.4)
7. (True—see section 4.5)
8. (False—see section 4.6)
9. (True—see section 4.7)
10. (True—see section 4.8)

References

1. Armstrong M (ed) (1995) Strategies for human resource management. A total business approach. Coopers & Lybrand, Kogan Page, London
2. Robbins SP (1991) Management, 3rd edn. Prentice-Hall, Englewood
3. Hellriegel D, Slocum JW (1993) Management, 6th edn. Addison-Wesley Publishing Company, Reading, Massachusetts
4. Anthony WP, Perrewé PL, Kacman KM (1996) Strategic human resource management, 2nd edn. The Dryden Press, Harcourt Brace College Publishers, USA
5. Taylor F (1911) Principles of scientific management. Harper & Row, New York
6. Gilbreth FB, Gilbreth LM (1914) Applied motion study. Sturgis and Walton, New York
7. Fayol H (1925) Administration industrielle et générale. Dunod, Paris
8. Weber M (1946) From Max Weber: essays in sociology. Oxford University Press, London
9. Hodgetts RM, Kroeck KG (1992) Personnel and human resource management. The Dryden Press, Harcourt Brace Jovanovich College Publishers, USA
10. Robbins SP (1991) Organizational behavior. Concepts, controversies and applications, 5th edn. Prentice-Hall International Editions, Englewood Cliffs
11. Luthans F (1995) Organizational behavior, 7th edn. McGraw-Hill, New York
12. Hendry C (1995) Human resource management: a strategic approach to employment. Contemporary Business Series, Butterworth-Heinemann Ltd, Oxford
13. Beaudwell I (2001) An introduction to human resource management: strategy, style or outcome. In: Beaudwell I, Holden L (eds) Human resource management: a contemporary approach, 3rd edn. Financial Times, Prentice-Hall, Harlow, pp 4–31
14. Mabey C, Salaman G, Storey J (eds) (2002) Strategic human resource management. Sage Publications, London
15. Carnegie D (1936) How to win friends and influence people. Harper & Row, New York
16. Maslow A (1943) A theory of human motivation. Psychol Rev 50:370–396
17. McGregor D (1960) The human side of enterprise. McGraw-Hill, New York

18. Drucker PF (1954) The practice of management. Harper & Row, New York
19. Herzberg F (1966) Work and nature of man. World, Cleveland
20. McClelland DC (1987) Human motivation. Cambridge University Press, Cambridge
21. Perrow C (1970) Organization analysis: a sociological view. Tavistock, London
22. Pfeffer J (1981) Power in organizations. Pitman, Marshifield
23. Liu M (1984) Approach socio-technique de l'organization. Les Éditions D'Organisation, Paris
24. Beaumont PB (1994) Human resource management. Key concepts and skills. Sage Publications, London
25. Brewster C (1994) European HRM. Reflection of, or challenge to, the American concept? In: Kirkbride PS (ed) Human resource management in Europe. Perspectives for the 1990s. Routledge, London, pp 56–89
26. Cardoso CC (1998) O conceito e as práticas de gestão de recursos humanos: evolução, perspectivas e controvérsias. *Série Documentos de Trabalho.* DT Nº 21/98 – GAP, Escola de Economia e Gestão, Universidade do Minho
27. Guest D (1987) Human resource management and industrial relations. J Manage Stud 24 (5):503
28. Torrington D (1998) Crisis and opportunity in HRM. The challenge for the personnel function. In: Sparrow P, Marchington M (eds) Human resource management. The new agenda. Financial Times, Pitman Publishing, London, pp 23–36
29. Martell K, Carroll SJ (1995) How strategic is HRM? Hum Res Manage 34(2):253–267
30. Budhwar PS (2000) Evaluating levels of strategic integration and development of human resource management in the UK. Pers Rev 29(2):141–157
31. Mabey C, Salaman G, Storey J (2002) Strategic human resource management: the theory of practice and the practice of theory. In: Mabey C, Salaman G, Storey J (eds) Strategic human resource management. Sage Publications, London, pp 1–13
32. Khatri NJ, Budhwar PS (2002) A study of strategic HR issues in an Asian context. Pers Rev 31 (1/2):166–188
33. Schuler RS (1992) Linking the people with the strategic needs of the business. Organ Dyn 4:18–32
34. Brewster C, Larsen HH (1992) Human resource management in Europe: evidence from ten countries. Int J Hum Res Manage 3(3):409–433
35. Brewster C, Larsen HH (1993) Human resources management in Europe: evidence from ten countries. In: Hegewisch A, Brewster C (eds) European development in human resource management. Kogan Page, The Cranfield Management Research Series, London, pp 126–148
36. Goshal S (2003) Liderança política e económica deve criar capacidade de sonhar. *Semanário Económico. Primeiro Plano.* (interview to Sumantra Goshal, by Luísa Rego and Maria Miguel Ferreira), 17 Apr 2003, pp 6–7
37. Shaeffer R (1989) Matching international business growth and international management development. Hum Res Plann 12(1):29–35
38. Noe RA, Hollenbeck JR, Gerhart B, Wright PM (2000) Human resource management: gaining a competitive advantage, 3rd edn. McGraw-Hill, International Editions, Boston
39. Fisher CD, Schoenfeldt LF, Shaw JB (1993) Human resource management, 2nd edn. Houghton Mifflin Company, International Student Edition, Boston
40. Mueller F (2002) Human resources as strategic assets: an evolutionary resource-based theory. In: Mabey C, Salaman G, Storey J (eds) Strategic human resource management. Sage Publications, London, pp 152–169
41. Huselid MA (2002) The impact of human resource management practices on turnover, productivity, and corporate financial performance. In: Mabey C, Salaman G, Storey J (eds) Strategic human resource management. Sage Publications, London, pp 104–127
42. Newton T, Findlay P (2002) Playing God? The performance of appraisal. In: Mabey C, Salaman G, Storey J (eds) Strategic human resource management. Sage Publications, London, pp 128–143
43. Storey J, Sisson K (2002) Performance-related pay. In: Mabey C, Salaman G, Storey J (eds) Strategic human resource management. Sage Publications, London, pp 144–151

44. Guest D (2002) Human resource management, trade unions and industrial relations. In: Mabey C, Salaman G, Storey J (eds) Strategic human resource management. Sage Publications, London, pp 237–250
45. Kovács I (1992) Novas tecnologias, recursos humanos, organização e competitividade. In: Kovács et al (eds) Sistemas flexíveis de produção e reorganização do trabalho. Lisboa, CESO I&D and PEDIP, pp 17–67
46. Bahrami H (2002) The emerging flexible organization: perspectives from Silicon Valley. In: Mabey C, Salaman G, Storey J (eds) Strategic human resource management. Sage Publications, London, pp 185–199
47. Atkinson J (1984) Manpower strategies for flexible organisations. Pers Manage 28–31
48. Atkinson J (1987) La flexibilité de l'emplois sur les marchés internes et externes de travail. In: Dahrendorf R, Kohler E, Piotet F (eds) Nouvelles formes de travail et activité. Dublin, Fundation Européenne pour l'Amélioration des Conditions de Vie et Travail, pp 3–31
49. Brewster C, Hegewisch A, Mayne L (1994) Flexible working practices. The controversy and the evidence. In: Brewster C, Hegewisch A (eds) Policy and practice in european human resource management. The Price Waterhouse Cranfield Survey, Routledge, London, pp 168–193
50. Brewster C, Mayne L, Tregaskis O (1997) Flexible working in Europe: a review of the evidence. Manage Int Rev, J Int Bus 37:85–103
51. Edwards P, Wright M (1998) HRM and commitment—a case study of team-working. In: Sparrow P, Marchington M (eds) Human resource management. The New Agenda. Financial Times, Pitman Publishing, London, pp 272–285
52. Wood S, Albanese MT (1995) Can we speak of high commitment management on the shopfloor? J Manage Stud 32(2):215–247
53. Solomon C (1989) The corporate response to workforce diversity. Pers J 42–53
54. Overman S (1991) Managing the diverse workforce. Hum Res Mag 36(4):32–36
55. Brewster C, Hegewisch A (1993) Personnel management in Europe: a continent of diversity. Pers Manage 25(1):36–40
56. Hopkins WE (1997) Ethical dimensions of diversity. Sage Publications (Sage Series on Business Ethics), Thousand Oaks
57. Barnes C (1992) Disability and employment. Pers Rev 21(6):55–73
58. Millward N, Woodland S (1995) Gender segregation and male/female wage differences. Gend Work Organ 2(3):125–139
59. Jacobs JA, Steinberg RJ (1995) Further evidence on compensation differentials and the gender gap in wages. In: Jacobs JA (ed) Gender inequality at work. Sage Publications, Thousand Oaks, pp 93–124
60. O'Leary VE, Ickovics JR (1992) Cracking the glass ceiling: overcoming isolation and alienation. In: Sekaran U, Leong FTL (eds) Womanpower. Managing in times of demographic turbulence. Sage Publications, Newbury Park, pp 7–30
61. Snizek WE, Neil CC (1992) Job characteristics, gender stereotypes and perceived gender discrimination in the workplace. Organ Stud 13(3):403–427
62. Jacobs JA (ed) (1995) Gender inequality at work. Sage Publications, Thousand Oaks
63. Stone P (1995) Assessing gender at work: evidence and issues. In: Jacobs JA (ed) Gender inequality at work. Sage Publications, Thousand Oaks, pp 408–423
64. York KM (1989) Defining sexual harassment in workplace: a policy-capturing approach. Acad Manage J 32(4):830–850
65. Miller P (2002) Strategy and the ethical management of human resources. In: Mabey C, Salaman G, Storey J (eds) Strategic human resource management. Sage Publications, London, pp 30–45
66. Hiltrop J, Janssens M (1990) Expatriation: challenges and recommendations. Eur Manage J 8 (1):19–26
67. Brewster C (1991) The management of expatriates. Kogan Page, London
68. Dowling P, Schuler R, Welch D (1994) International dimension of human resource management, 2nd edn. Wadsworth Publishing Company, California

69. Caligiuri P, Stroh L (1995) Multinational corporation management strategies and international human resource practices: bringing IHRM to the bottom line. Int J Hum Res Manage 6(3):494–507
70. Brewster C et al (1996) Comparative research in human resource management: a review and an example. Int J Hum Res Manage 7(3):585–604
71. Holden L (1996) HRM and employee involvement in Britain and Sweden: a comparative study. Int J Hum Res Manage 7(1):59–81
72. Jacques R (2002) Managing for the next century—or the last? In: Mabey C, Salaman G, Storey J (eds) Strategic human resource management. Sage Publications, London, pp 269–279
73. Sparrow P, Marchington M (1998) Re-engaging the HRM function. Rebuilding work, trust and voice. In: Sparrow P, Marchington M (eds) Human resource management. The new agenda. Financial Times, Pitman Publishing, London, pp 296–313
74. Mathis RL, Jackson JH (2008) Human resource management. South Western Cengage Learning, Mason
75. Melo P, Machado C (2015) Gestão de recursos humanos nas pequenas e médias empresas: contextos, métodos e aplicações. RhEditora, Lisboa
76. Carrel MR, Elbert NF, Hatfield RD (1995) Human resource management: global strategies for managing a diverse work force. Prentice Hall, London
77. Gomes JF, Cunha MP, Rego A, Cunha RC, Cabral-Cardoso C, Marques CA (2008) Manual de gestão de pessoas e capital humano. Edições Sílabo, Lisboa
78. Gómez-Mejia LR, Balkin DB, Cardy RL (1995) Managing human resources. Prentice-Hall, Englewood Cliffs, NJ
79. Anthony PA, Perrewé PL, Kacmar KM (1993) Strategic human resource management. The Dryden Press, Fort Worth, TX
80. Cardon SM, Stevens CE (2004) Managing human resources in small organizations: What do we know? Hum Res Manage Rev 14:295–323
81. De Ketele J-M, Chastrette M, Cros D, Mettelin P, Thomas J (1988) Guia do formador. Instituto Piaget, Lisboa
82. Kirkpatrick D (1998) Evaluation of training programs: the four levels. Berret-Koehler, San Francisco
83. Armstrong M (1984) A handbook of personnel management practice, 2ª edn. Kogan Page, London
84. Caetano A (1999) Avaliação de desempenho: metáforas, conceitos e práticas. RH Editora, Lisboa
85. Dessler G (1994) Human resource management, 6ª edn. Prentice Hall, Upper Saddle River
86. Cascio WF (2015) Managing human resources: productivity, quality of work life, profits, 10th edn. McGraw-Hill, NY
87. Carlson DS, Upton N, Seaman S (2006) The impact of human resource practices and compensation design on performance: an analysis of family-owned SMEs. J Small Bus Manage 44(4):531–543
88. Câmara PB, Guerra PB, Rodrigues JV (2007) Novo Humanator: Recursos humanos e sucesso empresarial. Publicações D. Quixote, Lisboa

Training and Development in Organizations: Start at the Beginning

Ana Paula Vieira Gomes Ferreira

Abstract This chapter discusses, broadly, the T&D management process in organizations. Following a brief description of the process, a particular emphasis is given to the first step: the **needs assessment** stage and its implications in the other T&D stages and human resource management (HRM) activities. Afterward, some guidelines and tools to conceive, plan, and implement the needs assessment stage will be suggested. The main objectives of this section are, as followed, threefold: (1) to give a global view of the T&D process and its main goals; (2) to help understand why needs assessment is a vital step in T&D; and (3) to suggest guidelines and examples to effectively implement a needs assessment in organizations.

1 Introduction

After having worked quite a few years around training and development (T&D) aspects (as trainer, as trainee, as T&D consultant, and finally as an academic), it is clear to me that T&D is paramount for individual employees and for their organizations. Indeed, the benefits of training inside organizational contexts are well documented (for a review see [1–3]).

However, T&D is too often seen as a "magic potion" for resolving every organizational problem or situation. As such, sometimes organizations have an "overdose" of training activities that can frustrate managers and their employees [4]. Also, and again too often, T&D is seen by organizational members (both employees and employers) as a waste of money and/or time.

T&D is mainly a tool—a powerful one—to enhance organizational performance, competitiveness, and economic growth. Both organizations and employees, as well

A.P.V.G. Ferreira (✉)
Department of Management, School of Economics and Management,
University of Minho, Campus Gualtar, 4710-057 Braga, Portugal
e-mail: aferreira@eeg.uminho.pt

C. Machado and J.P. Davim (eds.), *MBA*,
Management and Industrial Engineering, DOI 10.1007/978-3-319-28281-7_4

nations, need training: Workers need it to develop their careers or to get better compensations; organizations need skillful and efficient employees to improve performance and productivity, promote competitiveness, decrease absenteeism and turnover, as well as to improve client satisfaction; governments depend heavily on a skilled labor force with the capacity to learn, adapt, and master competitiveness in a globalized economy.

Yet, the benefits of T&D for organizations will only fulfill their potential when starting with a well thought process and taking into account the major purposes of the companies and their stakeholders. One of the major difficulties in organizations' T&D process is its beginning. Whatever the purpose of T&D in each company (emphasis on "each"), it is paramount to start with the "right foot." And, normally, that does not happen.

The impression of a "bad start," concerning the T&D process, struck me first as a trainer in professional training courses subsidized by European Funds, directed to help the development and sustainable growth of Portuguese small- and medium-sized enterprises (SMEs). The "first bad impression" faded a little when working in a private bank, where things seem to be much more strategic. But, then, the purposes were quite different, ... As a lecturer, however, I hear from my master degree students (most of them working in several different companies) that the beginning of the T&D process in their companies might not be, again, the most favorable way to get started... for all of the involved.

So, with this professional experience in mind, I will try to share some T&D concepts, tools, and processes, and especially some of my concerns regarding the need of a good start in the organizations' T&D process.

This chapter discusses, in general terms, the T&D management process in organizations. It starts by defining what T&D is and discussing its purposes and benefits for organizations and its members. It follows with a brief description of the process, and a particular emphasis will be given to the first step: the **needs assessment** stage and its implications in the other T&D stages and human resource management (HRM) activities. Afterward, some guidelines and tools to conceive, plan, and implement the needs assessment stage will be suggested.

The main objectives of this section are, as followed, threefold: (1) to give a global view of the T&D process and its main goals; (2) to help understand why needs assessment is a vital step in T&D; and (3) to provide (suggest ...) guidelines and examples of some tools to effectively implement a needs assessment in organizations.

2 Training and Development: An Overview

Training is often seen as a planned and systematic process of learning in the sense of acquiring, modifying, and/or developing knowledge, skills, and abilities (KSA) in order to achieve and/or improve the employees' performance in the current job and prepare them for an intended job. It is concerned with current productivity, whereas **development** can be seen as a "general enhancement and growth of an individual's skills and abilities through conscious and unconscious learning" [5, p. 1]. The major differences between training and development are at the level of time and scope. Training is a short-term process, using instruction to solve technical problems and so it is used for specific job-related purposes. Development is a "marathon": a long-term process, where organizations and their employees proactively seek competencies, skills and knowledge to resolve eventual future problems. It is related with the necessity to meet future organizational needs (Table 1).

Although the purposes of **training** and the ones of **development** are somewhat different, the process is the same. So, T&D in organizational context is an HRM tool that can be thought of as a set of planned, methodical activities designed to help the acquisition of knowledge, skills, attitudes, behaviors, and competencies that are needed, in some way, both for the companies and their members (e.g., [2, p. 77]) (Table 1).

T&D is currently well established as a systemic process that brings advantages at social, organizational, group, and individual levels. For instance, Aguinis and Kraiger [1] referred some studies where economic growth, inclusion of a country in a major economic block, or increased qualification of the workforce are benefits that resulted from T&D initiatives, whereas Cheung and Chan [6] showed the importance of training regarding countries' competitive positions. At the organizational level, Salas and Cannon-Bowers [7], Aguinis and Kraiger [1], and Park and Jacobs [8] all pointed out the results of T&D activities in organizational performance as much in productivity, profit, or safety, as well as in the decrease of errors and individual productivity. Ubeda-García et al. [9] showed that training policies seem to impact both objective and subjective measures of organizational performance.

At a microlevel—group and individual—training activities improve communication, performance, and task coordination within teams [1, 2, 10], can enhance positive attitudes toward the organization and the job [11–14], and seem to be able to increase employability within the firm [15]. In summary, to achieve success companies need to master their ability to train and develop in employees, ways to adapt to, and manage change. Here, T&D systems play a key role in strategic HRM, contributing to the competitiveness, and business's success.

Table 1 Differences between training and development

	Training	Development
Description	Purchasing behaviors, knowledge, framed in a function (performance)	Looking for complex results, less likely to objectify in terms of performance
Objectives	Minimize individual differences and standardize behaviors	Increasing variability of behavior
Process	Mechanical	Organic
	Emphasis on predictable responses	Leads to less predictable responses
Orientation	Toward work, task	Toward the person
Content	Competencies and attitudes needed to specific tasks	Concepts to stimulate analytical and critical skills
Time frame	Short term	Long term

2.1 The Training Process: A Brief Overview

Training as a set of activities that are put in place when managing people in organizational contexts, is an HRM practice that can also be thought of as a process comprising four stages: needs assessment, training design, training delivery, and training evaluation [12, p. 77].

The first stage, needs assessment, is a stage that will be of particular concern in this chapter. It is vital to define how training is going to be accomplished.

The training design stage should be thought of in terms of "learning objectives, trainee characteristics, current knowledge about learning processes, and practical considerations…" [16, p. 403]. This stage is built on data gathered in the needs assessment part. Plans, projects, and/or training programs are then elaborated, as well as specific goals are set for each projected activity. Efficiency criteria must be ensured in order to guide and control the resources, define methods, predict, and anticipate results from training. This stage must also consider where and when training activities should take place, and whom (trainers and trainees) will participate [17].

When implementing and delivering training everything previously planned should take place in the field. Here is the paramount to attend to (1) pre-training conditions (individual's characteristics, motivation for training; learning atmosphere; trainers quality, …); (2) training methods and learning strategies; and (3) post-training conditions (training evaluation, training transfer, …). (e.g., [1, 2, 7]).

The last, but not least, is training evaluation. According to Kirkpatrick's [18, 19] model, one of the most used evaluation frameworks, assessing the training process, implies four levels of the analysis. The first one, **reaction**, regards the trainees' opinions concerning the overall degree of satisfaction with the training conditions and trainers. The second level targets **learning** outcomes: the knowledge, competences, and attitudes that were acquired (or not). Next, **behavior** must also be assessed: it is important to know how knowledge, competences, and attitudes acquired during training activities are transferred to the workplace and interfere with performance requirements. Finally, training should also be evaluated according to **results** in terms of financial or operational impact on the organization.

But, let us focus on the needs assessment.

3 Training Needs Assessment

How do people define or describe training needs assessment (TNA)? This is something that a small questionnaire could help discover. So, when asking my Facebook friends (they are actually not that many), 20 of them came up with some answers…. (Thank you for the time!!!!).

Table 2 Examples of answers for TNA definition—focus on workers

	Definition
1	A collection of relevant information to see what training people need
2	Assessment process of the workers' training needs
4	Auscultation of the subjects and skills that workers need to do better and/or learn, upgrade
18	Organization's search in order to fit its training plan to the workers' needs and expectations
20	Inventory of training areas that allow a group of employees to acquire general and specific skills

Table 3 Examples of answers for TNA definition—focus on gaps

	Definition
5	Identification of employees' training needs by the employer
6	Diagnosis of organizational deficiencies at the level of employee training
8	Survey of skills/competencies deficit regarding the organizational strategy
16	Difference between the profile of the function and the person's profile
19	Identification of gaps in knowledge, skills, and behavior of workers, considered important for their performance and for organizational success

When asked about **what TNA means**, twelve pointed out the organization's workers as the main subject of the definition. Some examples can be seen in Table 2.

What was unexpected, especially considering that quite a few of these participants have some experience in management, is the number of answers indicating the prevalence of the workers' point of view. If you consider TNA as a tool, an organization's tool, to help achieve its goals, it is surprising that people with some kind of management responsibility still see TNA almost as an employee "wish/needs list" of training courses.

Actually, it should not be that surprising. Some studies point out that organizations still use employees' inquiries and pre-defined list of T&D sessions as some of the most frequent instruments of assessing training needs [20, 21].

Other participants in this small inquiry gave a more complete definition focusing on gaps and on the organization. Table 3 shows some cases.

3.1 Toward a Definition that Fits the Context

Employees and their employers surely need to understand that T&D is not only for the sake of individual employees' wants and desires, nor for choosing whatever consultants have on their "menu." T&D can really be used to motivate employees and/or to reward them, but first companies need to think first if they can afford using

T&D for these purposes all the time. Let us think about SMEs…. most of these companies struggle to survive. SME's owners and managers may need T&D to innovate, to expand, to learn how to do more and/or better in order to compete with other companies. In early stages of development, these organizations probably cannot use training resources to compensate the employee of the month! They need to use training in order to meet their survival and growth goals.

There is some evidence that the use of European Funds for T&D, for instance, was not adequately used for the initial purposes (e.g., [22]) at least in some countries. When working as a trainer for some Portuguese companies that applied for European Funds, I really had the impression that some trainees were there simply because they were paid and that some companies used the training programs they choose to implement or to accept from consultants very wrongly. Later, I saw some cases and heard of others that seem to fail the purposes of T&D, not due to a bad practice, but because of a deficient knowledge of the importance, usefulness, and techniques of the training process. Also, it seems to me that some people have certain myths and beliefs about T&D; one that I hear a lot is that "training is always good"! True… for the individual… not necessarily for the organization….

Sometimes organizations and its members seem to forget that T&D is an HRM tool that must be used with a purpose. Managing means organizing and coordinating a set of activities, with **limited resources** in order to achieve specific objectives. Having more or less resources means to apply them efficiently in order to be effective.

Hence, in order to help manage companies, TNA should be thought, **first**, as a tool to determine **whether training is the right solution for organizational and workplace problems**. TNA can be seen as a process of gathering data in order to establish what training needs exist in a particular organizational set and, then, help organizations to achieve their goals [4, 17, 23]. It is a detector of gaps that will help to identify skills/competency gaps by isolating the difference in and between current and future skills/competency.[1] TNA is a "diagnosis of what needs to be trained, for whom, and within what type of organizational system" [2, p. 80].

3.2 The TNA Process

The goal here is to identify the needs of each organization in order to respond to its goals and, **above all, screening which ones can effectively be addressed by training** activities [23, 24]. This stage is the first of the designing process of T&D's plans and programs, and it must comply a rigorous data collection on the company's current situation, its mission, and its resources.

[1]Source: http://workforceplanningtools.com.au/tools/training-needs-analysis/, accessed in 16/09/2015.

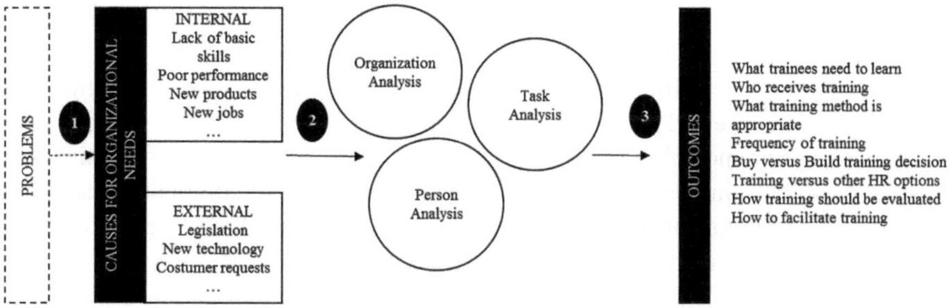

Fig. 1 Causes and outcomes of needs assessment [17, 25]

But, as a managing tool, we also need to go outside the organization. It is important to consider the context in hands. A scrutiny of the external environment, as well as of the internal, is required.

Let us see some examples:

- What challenges does the economic, legislative, and contractual environment where the company operates bring?
- What challenges are there in socio-demographics variables that might affect a company's human resources, as well as its clients?
- What are the industry and policy directions (national, state, local) that affect the company's business?
- Are there enough and good suppliers to operate with?

What we are describing here is the classical environment analysis that each organization needs to do. Ascertaining the external environment conditions is paramount to manage organizations and determine whether T&D activities are needed. Likewise, a company needs to "look inside":

- What is the strategic and business plan? Are there new projects, sites, facilities?
- Are there any problems attracting, recruiting, and retaining staff?
- How are the workforce productivity, quality, skills shortages, occupational health, and safety, risk?
- Is the organization a significant or a small employer?

As we could see above,[2] an external analysis as well as an internal is needed to give direction to an organization. This is also true for assessing the need of training. In Fig. 1, there are some examples of causes (internal and external) that trigger the need for a closer analysis inside the organization, and specifically for training purposes.

However, in order for T&D (or any other management tool) to be considered as a solution for something, a problem needs to be detected before searching the

[2]These, and other, examples can be seen in http://workforceplanningtools.com.au/tools/training-needs-analysis/.

causes. So, how is a problem detected in organizations? Well, just like a disease is ... by the symptoms and signs.

We really need to know the causes of something in order to reach a solution, and those causes are inferred by the signs that the various organizational systems give: absenteeism, unpunctuality, poor performance evaluations, increased mistakes, production failure, increased labor accidents, increased client complaints, increased turnover, introduction of a new system, process and/or machine, significant business opportunity that involves a major change, ...

These are the signs that are going, after careful analysis, to give companies the causes of the problems (Fig. 1, step ❶) or new situations and help to provide the adequate solutions.

Consider these examples:

Example 1

Bank ABC noticed an increase in complaints in customer service of a particular agency and notified these results to the branch manager, Mary. She asked the HRM Department to implement an attendance program for her employees to take part in.

An increase in complaints is one of the signs that help organizations see that something might not be right. Mary seems to think that an attendance course is the solution. Is she right?

Example 2

John is the head of the marketing department in an insurance company. As supervisor of a very conflicting employee, Mike, he saw that employee ruining the department's climate.... In the last 3 months, every single colleague presented a complaint about him: offensiveness, insolence, ... Apparently every time he hears a "no".... John sent Mike to a conflict management training program.

In this case, John believes that the solution for his conflicting employee is to send him to learn how to manage conflicts.... I wonder....

What we see in these examples are signs (number of complaints—clients or colleagues) that are not in conformity with what is generally expected: an organization does not want clients' complaints, and a team does not want a conflicting employee. So what to do with those signs? Mary saw the problem as being inside her staff: eventually a lack of competencies in attendance; John saw the root of the problem in a particular employee.

We need to understand why these signs appear before considering a solution, because we need to know the causes of these signs to set a proper solution. So, it is important to consider the context where those signs appear. The increase in clients' complaints may not be due to the inability of the staff.... It can be due to technology (the software is not appropriate, does not respond efficiently ...), to communication with other company's units, to shortage of staff .... If we know the causes, we will be better prepared to respond to the real problem... and the response may not be training. We can, eventually, respond with T&D activities for communication problems (depending on the issue...), but cannot solve staff scarcity with training... Right?!

Regarding the second example, we do not send a “trouble maker” to learn how to manage the conflicts he starts! That is not going to “put things right.” If the analysis of the problem is in Mike’s personality, we should know that personality is something that is not easily changed. So, John should consider other options besides T&D activities, like transfer him to another unit where Mike could work with a minimum of human interaction or even dismiss him.

Another thing we should be aware of is that people tend to confuse symptoms/facts with causes/opinions (Table 4). Consider the following examples:

- Employees serve customers poorly,
- Cannot get computer information,
- Salespeople are incompetent,
- There is a problem in attendance,
- The machine is broken,
- The deadlines are not met,
- The objectives are not being met, and
- The chief does not solve the problem.

Which ones are symptoms or facts and which ones are opinions or causes?

In the first column, we are basically saying that we know who to blame for a problem or situation. Our opinion is that the cause of the situation is the machine, the salespeople, the employees, and the chief.... It might be truth, really.... But we should consider all facts in order to discover the cause, since a machine may be ok, but we do not know out to work with it; salespeople and employees may be quite good, but they are not good enough, or even sufficient, to achieve the purposes; and the chief may not have enough authority to solve the problem.

This separation is important in order to find the right solution for an organizational problem, and **to know whether T&D activities are part of that solution**.

Normally T&D is a solution, or at least more efficient, when the cause of a problem/situation is related to the lack of knowledge and skills (Fig. 2).

Now that we were able to isolate the causes, we can consider if the solution depends on training or not. If so, an analysis at the organizational, task, and individual levels should be addressed by using a set of information-gathering instruments, and sources (Fig. 1, step ❷) ending with decision-making about where, when, and how to perform training policies and practices [17, 26, 27]. Table 5 shows some examples:

Table 4 Differences between causes/opinions and symptoms/facts

Causes/opinions	Symptoms/facts
Employees serve customers poorly	Cannot get computer information
Salespeople are incompetent	There is a problem in attendance
The machine is broken	The deadlines are not met
The chief does not solve the problem	The objectives are not being met

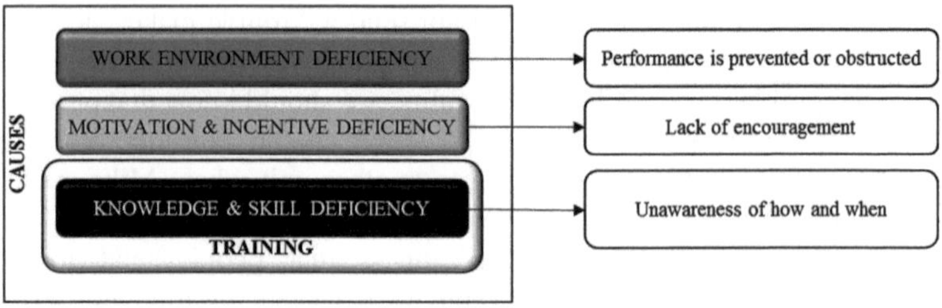

Fig. 2 Causes and effects [26]

Table 5 Sources of information [26, 28]

Levels	Sources
Organization	Organization's mission, goals, policies, business plans
	Organization chart
	Organizational climate indicators (absenteeism, job accidents, sickness leaves, organizational culture, …)
	HR indicators (quantity, competences and skills needed, …), exit interviews
	Performance indices (profitability, costs of labor and of materials, equipment usage, wastage, …)
	Requests for training (from managers and other organizational members)
	Quality
Task	Job and competencies descriptions (present and future)
	Skills, competencies, and performance standards
	Observation regarding the job
	Literature review (professional journals and magazines, professional associations, …)
	Interviews with job holders, supervisors, clients, suppliers
	Analyzing operational problems (waste, repairs, late deliveries, complaints, …)
Person	Individual performance indicators (productivity, accidents, quality, customer complaints, …) and comparison with standards
	Tests (job knowledge, skills, personality, …)
	Attitude surveys
	Performance appraisals
	Interviews
	Questionnaires
	Critical incidents

The needs assessment process, like many things in organizational life, is dependent on resources devoted to T&D activities in each company. So consequently, we need to know whether T&D activities fit with companies' strategic goals and if there is budget, time, and expertise and will to spend on these activities. So we need to answer 3 questions before starting:

1. What is the company's strategy?
2. Which resources are available?
3. Is there support and interest in T&D?

The first question is about identifying the level of involvement that T&D might have in helping to achieve the company's mission. Information about available resources will help us to choose the instruments and techniques. And finally, it is paramount to know the level of support employees, managers, and peers can give to T&D. Basically, these are the questions that need to be answered at an organizational level of analysis: The answers are going to determine the way the person responsible for performing the TNA will manage the process: the choice of resources, the time to perform it, the targets of T&D activities.... In sum, what to do in order to know: What? Why? When? Where? Who? How?

Task (job) analysis results in a description of the work activities of the employees. TNA can use the existing ones to achieve its purposes or help to build new ones or reformulate the existing.

Regarding person analysis, the purpose of it is to determine which employees need training in order to improve or achieve the wanted performance standards. For that, we need to know the quality of the human resources an organization has. This analysis requires understanding and identifying the person's characteristics regarding KSAs and attitudes, as well as the conditions for learning (social support, opportunity to perform, contextual constraints, feedback provided, ...) and the expected performance. Of no minor importance is understanding, the rewards and benefits T&D can provide for the employee.

4 Needs Assessment Techniques: Advantages and Disadvantages

The purposes of a TNA are achieved, as so many things in management, by collecting both qualitative and quantitative data for analysis. These data will help us to interpret current and future situations, problems, opportunities, and help to achieve what organizations want.

The key issue here is "personalization." As most things, T&D process should be "tailor made" for each organizational context. For instance, we cannot expect a SME to have the same human and financial resources to apply on T&D as bigger companies. This alone is one of the major constraints companies can have regarding T&D.

Much of the data is gathered by unique and customized instruments that need to be built for the purpose. We can obtain data using a questionnaire or interviews (or both...). But the reason why we are going to ask is because the company noted a sign or symptoms that caught its attention: An increase in turnover, absenteeism, lower productivity, or simply because a new machine or process is going to be implemented in a specific unit. So we need to shape the questions (about (non) existent knowledge

Table 6 TNA techniques [17, 25, 29–31]

Technique	Main advantages	Main disadvantages	Examples of use
Observation	Relevant data Low interference in work	Expert (observer) Employee behavior is affected Interpretation Observer dependency	Attendance, conflict, machine operation
Questionnaires	Inexpensive Amount of data in short term Objective, summarized data	Low return rates Not detailed Information restricted to the questions asked	Job satisfaction, motivation State of the employer/employee relationship
Interviews	Detailed information (causes; solutions) Allows exploring unexpected issues Flexibility (time, questions, …)	Time consuming Need deep analysis Expert (interviewer) Interviewer dependency	Job satisfaction, motivation Employer/employee relationship Improvement of processes Communication
Focus groups	Allows exploring unexpected issues Several perspectives at the same time Rapid to organize Relatively low cost	Time consuming Expert (interviewer) Eventual reluctance of participants (issues of confidentiality and different status among participants, …) Moderator dependency Social desirability bias	Complex and controversial issues (workplace bullying, …) Impact of new product
Documentation	Objective information about procedures Good information about job tasks	Technical language Obsolete information	Job and competencies analysis
Performance data	Objective information related to performance gaps (organization, unit, and individual levels)	Does not always explain the causes (just shows the symptoms)	Specific knowledge and behavior deficiencies Production problems
Tests	Objective information Easy and fast to apply	Not detailed Information restricted to the questions asked	Knowledge and behavior deficiencies

or skills, job satisfaction, conflicts, remuneration, relationship with supervisor …) in order to find out why those symptoms came up in that particular unit. The questions must also be made according to the target: language issues, level of literacy, organization and unit's goals, as well as employees' expectations and needs must be considered.

To perform an adequate TNA, collecting relevant data of what we want to measure is paramount. The type of information needed requires diverse instruments and methods. Collecting data regarding knowledge, attitudes, skills, or behaviors is different: We may need to use different methods, different techniques, different instruments….

The most common techniques to collect data for TNA and its (dis)advantages can be seen in Table 6.

Other techniques are available like brainstorming, and critical incident techniques. But besides collecting information, the analysis of that information is also needed. Analysis means, according to the online Oxford Dictionary,[3] a "detailed examination of the elements or structure of something" and a "process of separating something into its constituent elements." So, with the information in hand we need to see what the relevant parts are. Basically, we need to perform the three levels of analysis described before: organizational, task, and individual.

Time, resources, habits, and other technical and logistic considerations need to be considered in order to choose the methods to use in TNA. Because each method/technique has specific advantages and disadvantages more than one should be used and should be chosen considering the (dis)advantages of each: The disadvantages of one should be compensated by the advantages of the other.

5 Managing Needs Assessment: Some Issues

Roscoe [26] presents a very useful description of the needs assessment process and its management. In order to plan the TNA activity, the author signals the need to appoint a responsible person. This person, alone or with a team, needs to collect and analyze information to identify a problem, or to decide whether a training request is viable, by separating the causes from the effects. This alone will help to review and modify the initial ideas and proposals. With the information and its analysis in hands, the responsible team should generate and evaluate solutions that solve the causes. Some of the solutions may be T&D activities, others may not.

The key issue here is the choice of a solution. A set of possible solutions will surely be generated, and the choice may be problematic. A practical way to help choose a solution may be the consideration of some issues that can work as a check list for the person, or team, responsible for performing a TNA or managing T&D activities (Table 7).

This, of course, serves both for finding solutions in general to solve a problem or situation, and choosing, if the solution is T&D, which type of T&D activities can be performed in a particular organization.

[3]http://www.oxforddictionaries.com/pt/defini%C3%A7%C3%A3o/ingl%C3%AAs/analysis, accessed in 16/09/2015.

Table 7 Evaluating solutions [26]

Priorities	Criteria for evaluate	Reasons to reject
Legal requirements	Possibility of operating—will it work?	Overlap with other T&D activities
Business needs	Contribution to the situation—will it improve?	Lack of potential
Pay off	Budget—is it within the budget?	Inability to learn
Organizational politics	Costs and benefits	Costs
Staff motivations and expectations	Deadlines—will it be ready on time?	Time
Resources available	People involved—is it acceptable for everyone?	Culture of the organization
Management considerations	Resources—which ones are available?	Unavailability of resources

After deciding on the solution that better seems to solve the problem, a report must be presented for approval and, after acceptance, we need to implement it. Being a T&D solution accepted means that the proposal indicates a need of training (Fig. 1; step ❸) and answers were given to the outcomes presented in Fig. 1:

- What trainees need T&D activities?
- What are the best training methods?
- When and how many times T&D activities should occur to help resolve the situation?
- Are we going to “buy” or “build” the proposed T&D solution?
- Etc....

6 Final Considerations

TNA helps organizations to make sure that training is the appropriate solution for a problem, whether this problem is a performance gap, a promotion situation, or a socialization issue [4]. For this goal to be successfully achieved, we should consider, in my opinion, three main issues.

Personalization is one of them. Each organization is unique. Some have tradition in using T&D, others do not. Some have a culture of learning and innovation, and others have a culture of tradition. Some need very qualified employees, others need flexible employees.... So T&D must be used accordingly and the methods, instruments, and tools that we can use must be chosen and/or built to meet the needs and resources of each company. “Copy and paste” is not a guideline to use in T&D, and especially when performing TNA. We can be inspired by some new theories, methods, and instruments but we cannot always use them exactly as others do. We need to adapt, we need to question whether it serves the purposes, we need to see whether it deals with the issues we need to solve; we might need to add a new question in an interview or questionnaire, we might need to adjust the language, the length of the questionnaire....

Another issue is **managing**, managing in the sense formerly discussed: organizing and coordinating a set of activities, with limited resources in order to achieve specific objectives. The key words here are limited resources. Sometimes, we have a good budget and we can afford to use complex methods and instruments, hire an expert, and buy a new amazing knowledge test.... Sometimes, we do not. And we need to improvise, adapt, and think about how to use what we have.

Last, but not least, is **dedication**. According to Oxford Dictionary,[4] dedication means as follows: "the quality of being dedicated or committed to a task or purpose" and "the hard work and effort that somebody puts into an activity or a purpose because they think it is important." Do I need to say anything else?

7 Developing Skills

To exercise your knowledge, please consider the examples given in Sect. 3.2:

Example 1
Bank ABC noticed an increase in complaints in customer service of a particular agency and notified these results to the branch manager, Mary. She asked the HRM department to implement an attendance program for her employees to take part in.

Example 2
John is the head of the marketing department in an insurance company. As supervisor of a very conflicting employee, Mike, he saw that employee ruining the department's climate.... In the last 3 months, every single colleague presented a complaint about him: offensiveness, insolence, ... Apparently every time he hears a "no".... John sent Mike to a conflict management training program.

In order to establish if T&D activities are the solution for each of these cases, what would you do?

Please consider:

- The context: private big-/medium-sized companies (financial and insurance areas).
- Probable sources of information;
- Techniques for analysis;
- Organizational members involved.

---

[4]http://www.oxforddictionaries.com/pt/defini%C3%A7%C3%A3o/ingl%C3%AAs/dedication?searchDictCode=all, and http://www.oxforddictionaries.com/pt/defini%C3%A7%C3%A3o/learner/dedication, accessed in 16/09/2015.

References

1. Aguinis H, Kraiger K (2009) Benefits of training and development for individuals and teams, organizations, and society. Annu Rev Psychol 60:451–474
2. Salas E, Tannenbaum SI, Kraiger K, Smith-Jentsch KA (2012) The science of training and development in organizations: what matters in practice. Psychol Sci Public Interest 13(2): 74–101
3. Tharenou P (2010) Training and development in organizations. In: Wilkinson A, Bacon N, Redman T, Snell S (eds) The sage handbook of human resource management. Sage, London, pp 155–172
4. Cekada, T. (2011). Need training? Conducting an effective needs assessment. Prof Saf 28–35
5. Buckley R, Caple J (2000) The theory and practice of training. Kogan Page, London
6. Cheung HY, Chan A (2012) Increasing the competitive positions of countries through employee training. Int J Manpow 33(2):144–158
7. Salas E, Cannon-Bowers J (2001) The science of training: a decade of progress. Annu Rev Psychol 52:471–499
8. Park Y, Jacobs RL (2011) The influence of investment in workplace learning on learning outcomes and organizational performance. Hum Resour Dev Q 22(4):437–458
9. Ubeda-García M, Marco-Lajara D, Sabater-Sempere V, García-Lillo F (2013) Does training influence organizational performance? Analysis of the Spanish hotel sector. Eur J Train Dev 37(4):380–413
10. Kraiger K (2014) Looking back and looking forward: trends in raining and development research. Hum Resour Dev Q 25(4):401–408
11. Ehrhardt K, Miller J, Freeman S, Hom P (2011) An examination of the relationship between training comprehensiveness and organizational commitment: further exploration of training perceptions and employee attitudes. Hum Resour Dev Q 22(4):459–489
12. Vidal-Salazar M, Hurtado-Torres N, Matías-Reche F (2012) Training as a generator of employee capabilities. Int J Hum Resour Manage 23(13):2680–2697
13. Latif K (2012) An integrated model of training effectiveness and satisfaction with employee development interventions. Ind Commer Train 44(4):211–222
14. Pajo K, Coetzer A, Guenole N (2010) Formal development opportunities and withdrawal behaviors by employees in small and medium-sized entreprises. J Small Bus Manage 48 (3):281–301
15. Groot W, Van den Brink H (2000) Education, training and employability. Appl Econ 32: 573–581
16. Tannenbaum S, Yukl G (1992) Training and development in work organizations. Annu Rev Psychol 43:399–441
17. Noe R (2010) Employee training and development. McGraw-Hill, New York
18. Kirkpatrick D (1996) Great ideas revisited. Train Dev 554–559
19. Kirkpatrick D (1998) Evaluation of training programs: the four levels. Berret-Koehler, San Francisco
20. Mathews B, Ueno A, Kekale T, Repka M, Pereira Z, Silva G (2001) Quality training: needs and evaluation—findings from a European survey. Total Qual Manag 12(4):483–490
21. Ferreira A, Leite R (2014) A glimpse at the Portuguese employees' perceptions of training and development: what they get is what they see? Int J Appl Manag Sci Eng 1(1):1–16
22. Tomé E (2012) European social fund in Portugal: a complex question for human resource development. Eur J Train Dev 36(2/3):179–194
23. Brown J (2002) Training needs assessment: a must for developing an effective training program. Public Pers Manag 31(4):569–578
24. Noe R, Tews M (2008) Strategic training and development. In: Storey J, Wright PM, Ulrich D (eds) The Routledge companion to strategic human resource management. Routledge, pp 262–284
25. Meignant A (2003) A Gestão da Formação. D. Quixote, Lisboa

26. Roscoe J (1995) Analysis of organizational training needs. In: Truelove S (ed) The handbook of training and development. Blackwell, Cornwall, pp 50–78
27. van Eerde W, Tang S, Talbot G (2008) The mediating role of training utility in the relationship between training needs assessment and organizational effectiveness. Int J Hum Resour Manag 19(1):63–73
28. Ceitil M (2000) O papel da formação no desenvolvimento de novas competências. In: Caetano A, Vala J (eds) Gestão de Recursos Humanos: contextos, processos e técnicas. RH, Lisboa, pp 325–355
29. Morgan DL (1996) Focus groups. Annu Rev Sociol 22:129–152
30. Robinson N (1999) The use of focus group methodoloy—with selected examples from sexual health research. J Adv Nurs 29(4):905–913
31. Craig M (1995) Techniques for investigation. In: Truelove S (ed) The handbook of training and development. Blackwell, Cornwall, pp 1–27

Further Reading

1. Altschuld J, Watkins R (2014) Needs assessment: trends and a view toward the future. New Directions for Evaluation. Willey, USA
2. Anderson G (1994) A proactive model for training needs analysis. J Eur Ind Train 18(3):23–28
3. Bowman J, Wilson J (2008) Different roles, different perspectives: perceptions about the purpose of training needs analysis. Ind Commer Train 40(1):38–41
4. Carlisle J, Bhanugopan R, Fish A (2011) Training needs of nurses in public hospitals in Australia: review of current practices and future research agenda. J Eur Ind Train 35(7): 687–701
5. Denby S (2010) The importance of training needs analysis. Ind Commer Train 42(3):147–150
6. Ferreira R, Abbad G (2014) Avaliação de Necessidades de Treinamento no Trabalho: ensaio de um método prospectivo [Training needs assessment at work: a prospective method]. Revista Psicologia: Organizações e Trabalho 14(1):1–17
7. Meneses P, Zerbini T (2009) Levantamento de Necessidades de Treinamento: Reflexões Atuais. Análise 20(2):50–64
8. Shah H, Gopal R (2012) Training needs analysis for bus depot managers at GSRTC. Eur J Train Dev 36(5):527–543
9. Truelove S (1995) The handbook of training and development. Blackwell, Cornwall
10. http://workforceplanningtools.com.au/tools/training-needs-analysis/
11. http://www.businessballs.com
12. http://www.businessdictionary.com/
13. http://www.portaleducacao.com.br/administracao/artigos/32077/meios-para-o-levantamento-das-necessidades-de-treinamento-mlnt
14. http://www.sinfic.pt/SinficWeb/displayconteudo.do2?numero=24931
15. http://www.trainingzone.co.uk/
16. http://www.xperthr.com/how-to/how-to-conduct-a-training-needs-analysis/6716/
17. https://www.td.org

Part III
Speaking About Accounting and Finance

Accounting as an Information System

**Ana Alexandra Caria, Anabela Martins Silva,
Delfina Rosa Rocha Gomes
and Lídia Cristina Alves Morais Oliveira**

Abstract This chapter aims to provide, for all those that pursue a managerial career in the private industry, public sector, government, technological and engineering area, the "core" concepts, principles, rules and techniques of the accounting area. Positioning accounting within its broader social, economic and historical context, this chapter provides useful insights concerning the needs of different users of accounting information and provides a clear understanding of the differences and complementarities between financial and management accounting. In detail, the chapter defines accounting; analyses the evolution of accounting from mediaeval times to present; identifies the users of accounting information and their needs; identifies the objectives of financial reporting; identifies the qualities that make financial statements useful; defines the basic elements of financial statements—assets, liabilities, equity, income and expenses; provides an understanding of the accrual basis and going concern assumptions which underlie the preparation of financial statements; explains the nature and purpose of accounting standards; and introduces the financial statements that appear in a set of published accounts. To complement, the main concepts of management accounting and the distinctiveness and usefulness of the information it provides for decision-making are described.

1 Accounting in Context

1.1 Introduction

When we reflect broadly upon accounting and its roles in organizations and society, different dimensions can be highlighted. In fact, accounting may be seen as a technical practice alone or may be more broadly conceived as a social practice, with

A.A. Caria · A.M. Silva · D.R.R. Gomes (✉) · L.C.A.M. Oliveira
Department of Management, School of Economics and Management, University of Minho, Campus Gualtar, 4710-057 Braga, Portugal
e-mail: dgomes@eeg.uminho.pt

C. Machado and J.P. Davim (eds.), *MBA*,
Management and Industrial Engineering, DOI 10.1007/978-3-319-28281-7_5

implications for social and organizational functioning, that goes beyond the consideration of accounting as a neutral, if not benign, technical practice [1, p. 481]. Additionally, it is important to acknowledge that "accounting goes on at a more macrosociological level than is commonly assumed" and that accounting rules can be considered "not as features of particular organizations, but as properties of institutional domains, national societies, or now the evolving world" [1, p. 484; 2, p. 348, 354].

This broader nature of accounting allows us to identify different images of accounting [3]: accounting as an ideology; accounting as a language; accounting as a historical record; accounting as a current economic reality; accounting as an information system; accounting as a commodity; accounting as symbolic rituals; accounting as rationale; accounting as imagery; accounting as experimentation; and accounting as distortion. So, the question that emerges is: what is accounting?

1.2 What Is Accounting?

In a simplified way, accounting is the "language of business". Unless you can understand accounting, you will not understand business. But it is a very special and a living language [4], with its own symbols that allow to describe any business activity.

To fully understand the business, it is necessary to understand the language/concepts of accounting. The accounting language is strong and flexible enough to change as society changes, being determined by the context in which it operates (e.g. cultural, political, economic, technological and social factors). Therefore, accounting has been recognized as a social and institutional practice [5–7].

A broad definition of accounting that has stood the test of time is the one provided by the American Accounting Association (AAA). Accordingly, accounting is defined as "the process of identifying, measuring and communicating economic information to permit informed judgments and decisions by users of the information" [8].

This process comprises three phases: to identify economic events relevant to business; to count and measure; and to communicate the collected information in an aggregated way to interested users.

Accounting is not only about the production of accounting information (which comprises the functions of recording, classification and measurement), but also concerns the uses of the information produced (related to the following functions: information for decisions, management control and communication).

Accounting is an essential tool to support decision-making process and for accountability and reporting. The decisions are based on financial and non-financial information. Historically, accounting information has been financial, but accounting is increasingly being used to address the social, environmental, economic and governance concerns.

1.3 From Medieval Times to the Present

Although pervasive in all aspects of the public and private domains in our society, accounting has a long history. Several studies positioned the origins of bookkeeping in ancient civilizations, dating back to about 4000–3000 B.C., such as Babylonian, Assyrian and Sumerian civilizations, with some rudimentary forms of record-keeping [3, 9]. Bookkeeping went through significant changes with the Egyptian civilization, where the scribes had an important status in society; with the Greek Civilization, where the accounts of the Athenian State where made available to public scrutiny; and the Roman civilization, where laws were issued imposing the elaboration by taxpayers of statements of their financial positions [10]. The changes that bookkeeping went through the ancient world have "been attributed to various factors, including the invention of writing, the introduction of Arabic numerals and of the decimal system, the diffusion of knowledge of algebra, the presence of inexpensive writing material, the rise of literacy, and the existence of a standard medium of exchange" [3, p. 2; see also 10, 11].

The period of the Middle Age is considered as a period of stagnation for bookkeeping in the European context, and it is only around the fourteenth century that developments in bookkeeping happened and resulted in the use of double entry bookkeeping [10]. Previous to the development of double entry bookkeeping, the technique used to keep the accounting books was **single entry bookkeeping** or charge and discharge accounting. Charge and discharge accounting covers a wide range of accounting systems, which operate on a single entry [11]. As common features it can be highlighted "that they are accounts of an individual rather than an organization, [...] they cover the flow of resources over a period of time, and they make no distinction between capital and revenue transactions" [11, pp. 33–34]. As defined by Vangermeersch [12, p. 533], "Single entry bookkeeping is generally associated with accounting books containing only cash and personal (people and organizations) accounts (...)". The main book used was the **Memorial** where data were presented in a narrative or paragraph form, or presented in separate lines with figures in columns [11].

However, to meet with the growing business requirements, in particular the increasing level and complexity of trading operations, this method evolved with an increasing number of books being used and adapted, which resulted, at a certain point in time, in the **double entry bookkeeping system** [11]. This can be summarized as a system in which every transaction has a corresponding positive and negative entry (debits and credits) in a closed system of accounts, which include five categories: persons, values, cash, income and capital. This system has the following key features: (a) keeping of accounting records based on monetary measurement; (b) distinguishing between capital and income when analysing and classifying business transactions; and (c) the integrating role of the capital accounting, which is the recipient of all gains and losses, and the inherent state of balance between Assets and claims on those assets, also named Liabilities and Proprietorship [13, p. 311]. The characteristics of double entry bookkeeping can be summarized through the accounting equation: **Assets – Liabilities = Proprietorship or Owner's Equity**. As main books, the system uses the **Journal** and the **Ledger**, which contains accounts of each asset and liability of the business and of the amount invested (capital) of the owner [14, pp. 372–373]. To simplify the organization of the information, the different elements are structured in accounts, where an account comprises the elements of the patrimony of an entity with similar characteristics. By convention, it has been assumed the following structure of the account, representing the ledger book opened:

Debit	Title of the Account	Credit
Debits		Credits

As rules to register the movements in the accounts, it has been established, as a convention, the following:

	Debit	Credit
Asset	+	–
Liability	–	+
Revenue		+
Expense	+	
Capital	–	+

Note + means Increase in the patrimonial element; – means Decrease in the patrimonial element

To exemplify the use of both single and double entry bookkeeping, let us use the following example: The merchant bought chairs and a desk for 300 euros, paid in cash.

By using single entry bookkeeping, this transaction implies an expense and since it was paid the entry would be made only on the cash account: Decrease on the cash account.

Debit	Cash	Credit
		300 euros

By using double entry bookkeeping and for the accounting equation to be balanced, at least two accounts have to be used: Increase of Chair and Desk (Increase Asset) and Decrease in Cash (Decrease Asset).

Debit	Chairs and Desk	Credit	Debit	Cash	Credit
300 euros					300 euros

But where and how this method was developed and how it spread throughout Europe? It is accepted that the method emerged by the end of the Middle Age in the Northern Italian cities (Genoa, Florence and Venice) where Italian merchants were the leading businessmen of Europe, with a superior organization when compared with the other European countries [15, 16]. These Italian cities were until the fifteenth century the most developed in intellectual capability such as through education, writing, reading and arithmetic [17]. In fact, while some authors claim that the oldest records using double entry bookkeeping belonged to the Farolfi company (1299–1300) [18], others argue that the earliest records using double entry bookkeeping belonged to a municipality, the Massari of Genoa (1340) [19, 20].

The diffusion throughout Europe of double entry bookkeeping was stimulated in the fifteenth century by an important innovation—the printing press, and by the publication in 1494 of the first printed book on double entry bookkeeping Summa de Arithmetica, Geometria, Proportioni et Proportionalita written by Fra Luca Pacioli [11]. This invention made possible, and stimulated in the following centuries, the publication of a considerable number of accounting books that were widely diffused and accessible to a considerable number of literate and numerate people who previously did not have access to this instructional literature. In general, the vehicles of transfer used in the diffusion of double entry bookkeeping were books, foreign merchants and schools. Pacioli's pivotal book was an important instrument of diffusion, since it was translated into other languages and exerted a strong influence on the books published subsequently by authors from different countries [11, 21, 22]. However, the technique was not adopted on a large scale and many businessmen continued to routinely use simple, traditional accounting procedures they were accustomed to using. It was only after 1850 that double entry bookkeeping was adopted by a majority of businesses in Europe with the "manufacturing corporation, the income tax, and the emerging accounting profession as major stimulants" [23, p. 61]. The fact is that accounting theory and practice, in both financial and management accounting, either in private or in public institutions, went through considerable changes through the centuries, but double entry bookkeeping remains as the method still used in accounting today when preparing accounting information for different users, while single entry bookkeeping is only used for individuals and very small businesses.

1.4 Who Are the Users of Accounting Information and What Are Their Needs?

The accounting definition provided in Sect. 1.2 refers to the users of economic information and the decisions they make. Who are the users of accounting and why they need accounting information? The primary user of the accounting information is the business itself. In running a business, managers (e.g. finance directors, marketing managers and production supervisors) need past and prospective data for strategic and operational planning, for decision-making and for management control. Managers are named **internal users** of the accounting information. To these, information is accessible on an ongoing basis and its availability depends on the company itself.

In addition, there are individuals and organizations outside the entity who want financial information about the business. They are the so-called **external users** including present and potential investors, lenders, suppliers and other trade creditors, customers, competitors, employees, governments and their agencies, and the general public.

The information that each user needs depends upon the kinds of decisions the user makes. Table 1 illustrates some of the needs that accounting information may help to accomplish.

Though all the information needs of these users cannot possibly be met directly and on an ongoing basis, there are needs that are common to all users. These common needs are met through entity's preparation and presentation of general purpose financial statements (referred to as "financial statements").

Table 1 Users of accounting and needs for information of accounting users

Users	Types of needs. Need accounting information to …
Investors/shareholders	– Assess how effectively the managers are running the business – Assess the ability to pay dividends – Make decisions to buy, hold or sell ownership shares of a company
Lenders	– Evaluate the risks of granting credit or lending money – Assess the ability to pay interest and to repay the amount borrowed
Suppliers and other trade creditors	– Determine whether amounts owed to them will be paid when due. Trade creditors are likely to be interested in an entity over a shorter period than lenders unless they are dependent upon the continuation of the entity as a major customer
Customers	– Know if the company will continue to honour product warranties and support its product lines (especially when there is a long-term involvement or dependence on the entity)

(continued)

Table 1 (continued)

Users	Types of needs. Need accounting information to …
Competitors	– Compare performance
Employees	– Know about the stability and profitability of their employers – Assess the ability of the entity to provide remuneration, retirement benefits and employment opportunities
Governments, tax authorities, regulatory agencies	– Determine and assess taxation policies – Calculate national income and statistics – Decide whether financial support is needed (e.g. grants) – Tax authorities need to know whether the company complies (or not) with tax laws, and how much tax they should pay – Regulators need to know whether the company is operating (or not) within prescribed rules
General public	– Entities contribute to the local economy in many ways, including the number of people they employ and their patronage of local suppliers. Accounting information provides information about the trends and recent developments in the prosperity of the entity and the range of its activities

Source [24, §9] adapted

1.5 Financial Accounting and Management Accounting

There is only one discipline of accounting. As stated in Sect. 1.2, accounting identifies, measures and communicates financial information about economic entities to interested parties. However, a commonly distinction is made between financial accounting and management accounting, with the first being targeted primarily at those outside the business. Financial accounting (or general accounting, or financial accounting and reporting) is responsible for the preparation of financial reports on the entity for use by both internal and external users. While internal parties may require and have the power to obtain detailed information about the enterprise, most of the external users (e.g. investors, creditors, unions and government agencies) have to rely on the output of financial accounting—*financial statements* or *accounts*—as their major source of financial information. Financial statements form part of the process of financial reporting and are prepared and presented at least annually. The main components of a complete set of financial statements include the following, according to paragraph 10 of IAS 1:

- a statement of financial position as at the end of the period;
- a statement of profit or loss and other comprehensive income for the period;
- a statement of changes in equity for the period;
- a statement of cash flows for the period;
- notes, comprising significant accounting policies and other explanatory information [25].

On the other hand, management accounting (or internal accounting) covers the internal accounting of an organization. It is responsible for tailoring information that managers and decision-makers inside the company need to plan, control and evaluate an entity's operations. Management accounting identifies, measures, analyses and communicates to the managers information about resources that are used in the various business processes constituting the firm. Management accounting covers several areas: product cost issues, business-process cost analyses, budgeting, analysis of deviations from plans, management performance, strategic accounting and among others. Management accounting helps understanding how value is created so as to assist internal decision-makers.

2 Fundamentals of Financial Accounting

2.1 *Core Concepts*

Financial accounting is concerned with the identification, capture, record, process and presentation of the economic events (e.g. acquisition of resources, selling the firm's output, payment of the monthly wages, obtaining a loan) that take place between the company and the outside partners (e.g. suppliers, customers, tax authorities, and financial institutions). To allow the valuation of transactions, events have to be materialized by documents (e.g. invoices, receipts, bank statement, contracts, tax and social security filings) that comprise both financial and non-financial elements. Documents are chronologically recorded, classified and analysed so as to permit for the periodic construction of synthetic reports, the financial statements.

Financial accounting involves both the preparation and the use of information to facilitate economic decision-making of individuals and organizations. This goal is also enclosed in the objective of general-purpose financial reporting defined by the International Accounting Standards Board (IASB), the international accounting standards setter, in its Conceptual Framework for Financial Reporting.[1] Accordingly, the purpose of financial reporting is "to provide financial information about the reporting entity that is useful to existing and potential investors, lenders and other creditors in making decisions about providing resources to the entity" (Conceptual framework, §1.2) [26]. While a wide range of parties may be interested

[1] A conceptual framework establishes the concepts that underlie financial reporting. It is a coherent system of concepts that flow from the objective of financial reporting (level 1). The purpose of financial reporting provides guidance on identifying the qualitative characteristics that make accounting information useful and the elements of financial statements and assumptions (level 2). The third level identifies the recognition, measurement and disclosure concepts used in establishing and applying the accounting standards to achieve the financial reporting objective. The IASB's conceptual framework is described in the document, "Framework for Preparation and Presentation of Financial Statements". It was firstly issued in 1989 and revised in 2010 and 2015. The 2015s revision is expected to be completed in 2016. The analysis provided in this section refers to IASB Conceptual Framework [26].

in financial reporting, the IASB considers *decision usefulness* to capital providers (present and potential investors, lenders and other creditors) as the overarching objective.

The attributes that make the information provided in financial statements useful to users are called **fundamental qualitative characteristics**. If financial information is to be useful, it must be **relevant** and **faithfully represent** what it purports to represent (Fig. 1). Relevant accounting information is the one that is capable of making a difference in a decision. Information with no bearing on a decision is irrelevant. Financial information is capable of making a difference when it has predictive value, confirmatory value (it helps to confirm or revise their previous evaluations), or both (Conceptual framework, §§2.6–2.10) [26]. The second fundamental quality of accounting information is faithful representation. Faithful representation means that the numbers and descriptions match what really existed or happened. Faithful representation is necessary because most users have neither the time nor the expertise to evaluate the factual content of the information. Faithful representation represents the substance of an economic phenomenon rather than merely representing its legal form. To be a faithful representation, information must be (1) *complete* (it includes all information that is necessary for faithful representation), (2) *neutral* (free of bias), and (3) *free of material error*. For information to be neutral, it has to be *prudent*. Prudence is defined as the exercise of caution when making judgements under conditions of uncertainty (Conceptual framework, §§2.14–2.19) [26].

Adding to the fundamental qualitative characteristics, there are also the **enhancing characteristics** of useful financial information. These are characteristics that are complementary to the fundamental qualitative characteristics and help distinguish, for decision-making purpose, more useful information from less useful

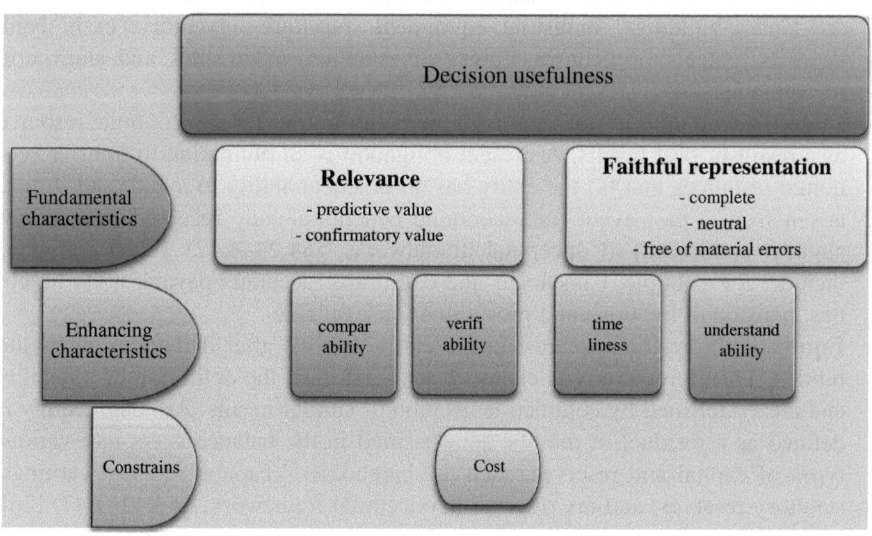

Fig. 1 Qualitative characteristics of financial reporting information

information. Enhancing characteristics of accounting information are given as: (1) *comparability* (to be useful, information about an entity has to be comparable, with similar information about other entities and with similar information about the same entity for some other period), (2) *verifiability* (different knowledgeable and independent observers could reach general consensus about the economic phenomena that is represented), (3) *timeliness* (to influence decisions information has to be available before it loses its opportuneness) and (4) *understandability* (users, who have a reasonable knowledge of business and economic activities and are able to read a financial report, comprehend the meaning of the information) (Conceptual framework, §§2.22–2.37) [26].

In providing information, companies must consider an overriding factor that limits the reporting. This constrain is cost. Cost constraint means that companies must weigh the costs of providing the information against the benefits that can be derived from using it. The benefits perceived to be derived from information must exceed the costs perceived to be associated with its preparation (Conceptual framework, §§2.38–2.42) [26].

Financial statements portray the financial effects of transactions and other events by grouping them into broad classes according to their economic characteristics. These basic building blocks are called the **elements of financial statements**. Assets, liabilities and equity are the elements of the statement of financial position and are associated with the measurement of an entity's financial position (usually called balance sheet). Income and expenses are the building blocks of the statement of profit or loss and are related to the measurement of performance. The definitions presented in the 2015 IASB Conceptual Framework are as follows:

- **Assets** are present economic resources controlled by the entity as a result of past events. An economic resource is a right that has the potential to produce economic benefits (Conceptual framework, §§4.5–4.23) [26]. Examples of assets are lands, buildings, industrial equipment, furniture, inventory, cash, bank deposits, patents or licenses, equity investments, receivables and short-term investments.
- **Liabilities** are present obligations of the entity to transfer an economic resource as a result of past events. A present obligation is an obligation to transfer economic resources that (a) the entity has no practical ability to avoid; and (b) has arisen from a past event (i.e. economic benefits already received or activities already conducted) (Conceptual framework, §§4.24–4.42) [26]. Liabilities include, for example, bank loans and overdrafts, accounts payables, tax liabilities, pension obligations and provisions for legal risks.
- **Equity** is the residual interest in the entity's assets after deducting all its liabilities. Therefore, equity is claims that do not meet the definition of a liability and are established by contract, legislation or similar means. Although equity is defined as a residual, it may be subclassified in the balance sheet into various types of capital and reserves, such as shareholders' capital, retained earnings, statutory reserves, and tax reserves (Conceptual framework, §§4.43–4.47) [26].

- **Income** is increases in assets or decreases in liabilities that result in increases in equity, other than those relating to contributions from holders of equity claims (Conceptual framework, §4.48) [26]. Sales, services rendered, assets yielding interests, royalties, and dividends are examples of incomes.
- **Expenses** are decreases in assets or increases in liabilities that result in decreases in equity, other than those relating to distributions to holders of equity claims (Conceptual framework, §4.49) [26]. Examples of expenses are the cost of goods sold, salaries and social charges, utility costs (electricity, water and telephone), advertising costs, interest expenses, insurance premiums, and depreciation and amortization of certain assets.

Though income and expenses are defined in terms of changes in assets and liabilities, information about income and expenses is just as important as the information provided by assets and liabilities (Conceptual framework, §4.52) [26].

Only items that meet the definition of an asset, a liability or equity are recognized in the statement of financial position and only items that meet the definition of income or expenses are to be recognized in the statement(s) of financial performance. **Recognition** involves depicting items in words and by a monetary amount, and including that amount in totals in the relevant statement. Nevertheless, recognition depends on three criteria: their recognition provides users of financial statements with (1) relevant information about the asset or the liability and about any income, expenses or changes in equity, (2) a faithful representation of the asset or the liability and of any income, expenses or changes in equity, and (3) information that results in benefits exceeding the cost of providing that information (Conceptual framework, §§5.9–5.24) [26].

The prevailing objective of financial reporting is providing information that is decision usefulness to a large number of users. Consequently, there are several accounting principles (also called assumptions, conventions and concepts) which underpin the preparation of the financial statements. These could be behavioural rules or very operational guidelines about practice. There are two overriding assumptions underlying financial statements. These are **the going concern** and **the accrual basis** assumptions.

The financial statements are normally prepared assuming the entity is a going concern and will continue in operation for the foreseeable future, with neither the intention nor the necessity of liquidation, ceasing trading or seeking protection from creditors pursuant to laws or regulations (Conceptual framework, §3.10) [26]. The assessment of an entity's ability to continue as a going concern is responsibility of the entity's management. This involves making a judgement, at a particular point in time, taking into account "all available information about the future which is at least, but is not limited to, twelve months from the end of the reporting period" (IAS 1, §26) [25]. If management has significant concerns about the entity's ability to continue as a going concern, the uncertainties must be disclosed. If management concludes that the entity is not a going concern, then the financial statements should not be prepared on a going concern basis, in which case IAS 1 requires disclosure the reasons why and the basis on which it prepared the financial statements (IAS 1, §25) [25].

The other underlying assumption is the accrual basis. Under this basis, the effects of transactions and other events are recognized when they occur and not when cash or its equivalent is received or paid. Therefore, transactions and other events are recorded in the accounting records and reported in the financial statements of the periods to which they relate. Using the accrual basis to determine net income means that a company recognizes income when it provides the goods or services rather than when it receives cash or its equivalent. Similarly, it recognizes expenses when it incurs them rather than when it pays them. For example, an energy bill owing at the accounting year end is thus treated as an expense for this year even if it is paid in the next year. In other words, net income must include all and only the income/expenses that have been earned/consumed during the accounting period. IAS 1 in paragraph 27 defines that "an entity shall prepare its financial statements, except for cash flow information, using the accrual basis of accounting" [25]. Financial statements prepared on an accrual basis better informs about the entity's present and continuing ability to generate favourable cash flows than does information limited to the financial effects of cash receipts and payments.

To a large extent, financial reports are based on estimates, judgements and models rather than exact representations. Measurement systems are fundamental to the determination of net income and to the measurement of net assets. Considering the objective of financial reporting, the qualitative characteristics of useful financial information and the cost constraint, there are different measurement bases for different assets, liabilities and items of income and expenses.[2] **Measurement** "is the process of quantifying, in monetary terms, information about an entity's assets, liabilities, equity, income and expenses" (Conceptual framework, §6.2) [26]. Measurement bases discussed in the 2015 IASB Conceptual Framework include the **historical cost** and **current value bases** (§6.4).

The measurement basis most commonly adopted by entities in preparing their financial statements is historical cost, which is usually combined with other measurement bases (e.g. inventories are usually carried at the lower of cost and net realizable value). Under the historical cost, assets are recorded at the amount of cash (or cash equivalents) paid to acquire them. Liabilities are recorded at the amount of proceeds received in exchange for the obligation, or in some circumstances (for example, income taxes), at the amounts of cash or cash equivalents expected to be paid to satisfy its liability. Although easy to use, historical cost measures of assets or liabilities do not reflect changes in prices. However, we do not use historical cost comprehensively in financial statements today. Measures based on historical cost do reflect changes such as the *consumption or impairment of assets* and the fulfilment of liabilities. This means that the historical cost of non-financial assets (e.g. property, equipment and patents) is adjusted over time to depict the depreciation or

[2]Currently, financial statement amounts are determined using a variety of measurement bases. Historical cost is used, for example, for cash. Impaired amortised historical cost is used for purchased property, plant, and equipment. Amortised historical cost is used for loans, receivables and long-term debt. Fair value is used for investment securities and derivatives. Entity-specific value is used for impaired inventories and impaired property, plant, and equipment.

amortization of the assets (consumption), and/or the fact that part of the historical cost of the asset is no longer recoverable (impairment). Historical cost base also includes the **amortised cost**,[3] the **deemed cost**[4] and the **current cost**.[5]

In addition to measures based on historical cost, there are current value measurement bases, which include (1) **fair value**, and (2) **value in use** for assets and **fulfilment value** for liabilities. These provide monetary information about assets, liabilities, income and expenses using information that is updated to reflect conditions at the measurement date. Therefore, current values reflect any positive or negative changes, since the previous measurement date, in estimates of cash flows and other factors included in those current values (Conceptual framework, §§6.19–6.20) [26].

Fair value is determined from the perspective of market participants (e.g. estimates of future cash flows, uncertainty inherent in the cash flows and the time value of money) and is defined as "the price that would be received to sell an asset, or paid to transfer a liability, in an orderly transaction between market participants at the measurement date" (Conceptual framework, §6.21) [26]. Fair value measurement is simple and verifiable if fair values can be observed in active markets. When it is not the case, valuation techniques (that can include the use of cash-flow-based measurements) may be needed to estimate that fair value. Consequently, identical assets or liabilities may be measured at different amounts.[6] IFRS 13-*Fair Value Measurement* sets out a framework for measuring fair value and requires disclosures about fair value measurements [13]. To increase consistency and comparability in fair

[3]The amortised cost is defined in paragraph 6.9 of IASB's Conceptual framework as following: "The historical cost of a financial asset is initially the value of the consideration given to acquire the asset plus the transaction costs relating to the acquisition. The historical cost of a financial liability (again, sometimes referred to as amortised cost) is initially the value of the consideration received to take on the liability less the transaction costs incurred in taking it on. The subsequent carrying amount of financial assets and financial liabilities measured using amortised cost reflects subsequent changes such as the accrual of interest, changes in the estimates of cash flows (including the impairment of financial assets) and payments or receipts, but does not reflect subsequent changes in prices caused by other factors" [26].

[4]"Assets acquired and the liabilities incurred in transactions that involve no exchange do not have a readily identifiable initial cost. In such cases, current values are sometimes used as a proxy for cost (deemed cost) on initial measurement and that deemed cost is then used as a starting point for subsequent measurement" (Conceptual framework, §6.11) [26].

[5]When price changes are significant, information about the current cost may sometimes be more relevant than information about their historical cost. "The current cost of an asset (liability) is the cost of (proceeds from) an equivalent asset (liability) at the measurement date. Current cost and historical cost are both entry values (i.e. they reflect values in the market in which the entity acquires the asset or incurs the liability)" (Conceptual framework, §6.18) [26].

[6]Because fair value measurement incorporates market information into the financial statements is the most relevant measure for financial instruments in the dominant market environment. Nonetheless, fair value measurement can pose challenges such as the absence of active markets, difficulties to determine fair value estimates in the absence of listed prices, and market illiquidity, such as the one that emerged during the recent credit crunch. For an analysis of the relevance of fair value accounting, see [27, 28].

value measurements and related disclosures, IFRS 13 establishes a fair value hierarchy that categorizes into three levels the inputs to valuation techniques used to measure fair value. Level 1 inputs are quoted prices (unadjusted) in active markets for identical assets or liabilities that the entity can access at the measurement date. Level 2 inputs are inputs other than quoted prices included within Level 1 that are observable for the asset or liability, either directly or indirectly. Level 3 inputs are unobservable inputs for the asset or liability (IFRS 13, §§72–90) [29].

Contrasting to fair value, value in use and fulfilment value are determined from the entity perspective. They are defined as the present value of the cash flows that an entity expects: (1) to derive from the continuing use of an asset and from its ultimate disposal, in the case of the value in use; (2) to incur as it fulfils a liability, in the case of the fulfilment value. Value in use and fulfilment value cannot be directly observed on the market, so they are determined using cash-flow-based measurement techniques (Conceptual framework, §§6.34–6.46) [26].

To select a measurement basis, the qualitative characteristics of useful financial information should be taken into consideration. Therefore, for information provided by a particular measurement basis to be useful, it must be relevant (e.g. what information it will be produced about both financial position and financial performance; how the asset or liability contributes to future cash flows; what are the characteristics of the asset or liability; what is the level of measurement uncertainty) and it must faithfully represent what it purports to represent (e.g. there should be a consistent use of measures for related items). In addition, the information provided should be comparable (that is, using the same measurement bases between periods and between entities), verifiable (implies using measurement bases that result in measures that can be independently corroborated), timely (this characteristic has no implication in the chosen measurement base) and understandable (depend on the number of different measurement bases used and on whether they change over time). In addition, as with all other areas of financial reporting, the cost constraint affects the selection of a measurement basis (Conceptual framework, §§6.53–6.63) [26]. The accounting standards, analysed in the following section, should define the appropriate measurement bases for each element of the financial statements.

2.2 Nature and Purpose of Accounting Standards

The users of financial statements, identified in Sect. 1.4, need to understand financial information in the same way. Accordingly, financial accounting is based on a set of rules and accounting standards. Accounting standards "include specific principles, bases, conventions, rules and practices necessary to capture the data and prepare the financial statements" [4, p. 15]. Originally, due to different political economic, social and cultural backgrounds, each country, through its regulatory bodies and/or professional accountancy bodies, has issued its own accounting standards or financial reporting standards.

However, since the middle of the twentieth century, the scenario of businesses has become increasingly global. Therefore, financial reporting is taking place in an international context that is characterized by impressive growth in international trade and foreign direct investment; cross-border mergers and acquisitions; the rise of multinational corporations; the widespread ownership of modern corporations; changes in world politics; an unstable international monetary system; and global international capital markets [30, 31]. Within this global scenario, "narrowly national views of accounting and financial reporting can no longer be sustained" [30, p. 5].

Pressures for common accounting standards—that is for **accounting harmonization**—have come from users, preparers and regulators of financial statements. But it was the globalization of the capital markets that brought to the forefront the increasing need for more comparable and reliable financial information. The major player promoting international accounting harmonization nowadays is the IASB, whose origins go back to the formation of its predecessor, the International Accounting Standards Committee (IASC) in 1973.

The IASB is the standard-setting body of the IFRS Foundation that aims to "develop a single set of high-quality, understandable, enforceable and globally accepted financial reporting standards based upon clearly articulated principles" [32]. Since its initial standards[7] characterized as insufficiently prescriptive, and that often permitted the use of almost all the accounting treatments existing in the world, to its current International Financial Reporting Standards (IFRS), the IASB has come a long way. IFRS are, nowadays, followed by overseas registrants in most of the world's stock exchanges, supported by international organizations (e.g. EU, G20, World Bank, IMF, IOSCO, Basel Committee), national accounting standard setters, governments, developing and emerging countries. Today, almost 120 countries require or permit the use of IFRS by public companies [32]. IFRS are a generally principles-based set of standards that covers all accounting and financial reporting topics. IFRS prescribe (1) the items that should be recognized as assets, liabilities, income and expense; (2) how to measure those items; (3) how to present them in a set of financial statements; (4) and related disclosures about those items. Table 2 presents the standards in force as on August 2015.

As stated, the IASB is an independent, private sector standards setter that has no authority to mandate or supervise the adoption of IFRS. Countries need to establish their own mechanisms for bringing IFRS formally into national law and for ensuring consistent and rigorous application. For the increased legitimacy and stature of the IASB worldwide, much has contributed the European Union's "Regulation No 1606/2002 on the application of International Accounting Standards" (known as IAS Regulation) that requires European companies, listed on a regulated market, to prepare consolidated accounts in accordance with the IASB

[7]IASB initial standards are called International Accounting Standards (IAS). IAS were issued by the IASC, predecessor of the IASB till 2000.

Table 2 IAS and IFRS as of 1 August 2015 (in brackets, the year of the original issue or major amendment)

IAS	IFRS
IAS 1—Presentation of Financial Statements (2003)	Preface
IAS 2—Inventories (2003)	Framework (2010)
IAS 7—Statement of Cash Flow (1992)	IFRS 1—First-time Adoption of International Financial Reporting Standards (2003)
IAS 8—Accounting Policies, Changes in Accounting Estimates and Errors (2003)	IFRS 2—Share-based Payment (2004)
IAS 10—Events after the Reporting Period (2003)	IFRS 3—Business Combinations (2004)
IAS 12—Income Taxes (1996)	IFRS 4—Insurance Contracts (2004)
IAS 16—Property, Plant and Equipment (2003)	IFRS 5—Non-current Assets Held for Sale and Discontinued Operations (2004)
IAS 17—Leases (2003)	IFRS 6—Exploration for and Evaluation of Mineral Resources (2006)
IAS 19—Employee Benefits (2011)	IFRS 7—Financial Instruments: Disclosures (2005)
IAS 20—Accounting for Government Grants and Disclosure of Government Assistance (2008)	IFRS 8—Operating Segments (2008)
IAS 21—The Effects of Changes in Foreign Exchange Rates (2003)	IFRS 9—Financial Instruments (2014)
IAS 23—Borrowing Costs (2007)	IFRS 10—Consolidated Financial Statements (2011)
IAS 24—Related Party Disclosure (2003)	IFRS 11—Joint Arrangements (2011)
IAS 26—Accounting and Reporting by Retirement Benefit Plans (1987)	IFRS 12—Disclosure of Interest in Other Entities (2011)
IAS 27—Separate Financial Statements (2011)	IFRS 13—Fair Value Measurement (2011)
IAS 28—Investments in Associates and Joint Ventures (2011)	IFRS 14—Regulatory Deferral Accounts (2014)
IAS 29—Financial Reporting in Hyperinflationary Economies (2008)	IFRS 15—Revenue from Contracts with Customers (2014)
IAS 32—Financial Instruments: Presentation (2003)	
IAS 33—Earnings per Share (2003)	
IAS 34—Interim Financial Reporting (1998)	IFRS for SMEs—The International Financial Reporting Standard for Small- and Medium-sized Entities (2009)
IAS 36—Impairment of Assets (2004)	
IAS 37—Provisions, Contingent Liabilities and Contingent Assets (1998)	
IAS 38—Intangible Assets (2004)	
IAS 40—Investment Property (2003)	
IAS 41—Agriculture (2008)	

Source http://www.ifrs.org/IFRSs/Pages/IFRS.aspx, accessed on 31 August 2015

standards, from 2005 onwards [33].[8] The IAS Regulation gives member states the option to require or permit IFRS as adopted by the EU in separate company financial statements (statutory accounts) and/or in the financial statements of companies whose securities do not trade on a regulated securities market. The approval of this Regulation is responsible for a radical change in the financial reporting and framework in most of the European countries and beyond.[9]

Despite the widespread global adoption of IFRS, international differences in corporate financial reporting remain. IFRS accounts are still influenced by pre-IFRS national accounting and reporting traditions. "This is partly because IFRS is used in many countries only for consolidated statements, and partly because different national versions of IFRS practice exist" [30, p. 19].[10] A profile about the use of IFRS in individual jurisdictions (the G20 jurisdictions plus 120 others) is available at http://go.ifrs.org/global-standards.[11]

2.3 *Accounting Process and Financial Statements*

Financial statements are the final output of the accounting process. The accounting process consists in a structured multiset progressive classification and aggregation process of elemental data [4]. This relies on the use of technical tools and requires human intervention and interpretation. The accounting process includes four stages: analysis of supporting documents; posting to accounts; control; and synthesis [4].

---

[8]In EU countries, IASB standards and interpretations are adopted in the form of Regulation and published in the Official Journal of the European Union (in all the official languages of the UE). Before becoming law in the EU, IFRS must go through due process of endorsement developed by the European Financial Reporting Advisory Group (EFRAG). The EU Regulations adopting IFRS are available at: http://ec.europa.eu/finance/accounting/legal_framework/regulations_adopting_ias/original_text_en.htm.

[9]In 2002, Australia and New Zealand also endorsed the adoption of IASB standards, starting from 2005 and 2007, respectively. That same year, the IASB and the Financial Accounting Standards Board (FASB), the regulatory body of the USA, jointly issued a memorandum of understanding, which formally stated their commitment to converge the standards from both organizations. In 2010, the two Boards started its policy of phasing in adoption of new major standards over several years.

[10]The IFRS regime ("principle based") allows for a degree of discretion that may led to inconsistent application of the standards across countries. Differences in the implementations of IFRS, different translations of IFRS, the optional treatments allowed under IFRS (the existence of overt option, and/or, covert options, vague criteria and interpretations in IFRS), the use of estimations in IFRS, are among the factors that explain that differences of practice still exist within IFRS usage [34, 35].

[11]Eighty-three per cent of the jurisdictions profiles (116 jurisdictions) require IFRS for all or most domestic publicly accountable entities (listed companies and financial institutions). The remaining 24 jurisdictions that do not yet require IFRS for all or most domestic listed companies already permit IFRS for at least some domestic listed companies.

In the stage of analysis of supporting documents, the document materializes each accounting transaction and is the basis for an entry in a journal. This accounting tool provides a chronological list of all transactions and it depends on the computer software.

Then the effects of the accounting events are transferred to (posted to) the specialized ledgers and the general ledger where the components of the description of the transaction are structured by nature and/or type of transaction. The posting to accounts' stage is a purely mechanical task.

The third stage of accounting process is control. Detailed individual accounts entries are grouped together and replaced by the balance of the aggregated account, and recapitulated in the trial balance. The aim is to check that the sum of all the debit entries or balances is equal to the sum of all the credit entries or balances.

Finally, the synthesis of the accounting process is materialized in financial statements, also called "accounts". According to IAS 1, paragraph 36, "an entity shall present a complete set of financial statements (including comparative information) at least annually. (…) Normally, an entity consistently prepares financial statements for a one-year period" [25]. Usually that date coincides with the end of civil or fiscal year. As reported in Sect. 1.5, the balance sheet, the profit or loss statement, the notes to financial statements, the statement of changes in equity and the cash flows statement constitute a complete set of financial statements. The first three documents are the minimum reporting required by a large majority of countries.

Balance sheet

The balance sheet reports on a company's financial position at a point in time. Assets, liabilities and equity are the elements of the balance sheet, which shows the accounting equation or balance sheet equation:

This equation keeps balanced due to the double entry accounting.

Assets are divided into two categories: **current** and **non-current assets**. According to IAS 1, paragraph 66 [25], "an entity shall classify an asset as current when: (a) it expects to realise the asset, or intends to sell or consume it, in its normal operating cycle; (b) it holds the asset primarily for the purpose of trading; c) it expects to realise the asset within twelve months after the reporting period; or (d) the asset is cash or cash equivalent (…) unless the asset is restricted from being exchanged or used to settle a liability for at least twelve months after the reporting period. An entity shall classify other assets as non-current".

The main categories of current assets are inventories (e.g. merchandise, raw materials, work-in-process and finished products), receivables (accounts receivable) and cash. Non-current assets are divided into tangible assets (e.g. property, plant and equipment), intangible assets (e.g. trademarks, patents or software) and financial assets (e.g. shares or bonds of another business, or the representation of medium- or long-term credit extended to a third-party business).

As assets, liabilities are also divided into **current** and **non-current liabilities**. According to IAS 1, paragraph 69 [25], "an entity shall classify a liability as current when: (a) it expects to settle the liability in its normal operating cycle; (b) it holds the liability primarily for the purpose of trading; c) the liability is due to be settled within twelve months after the reporting period; or (d) it does not have an unconditional right to defer settlement of the liability for at least twelve months after the reporting period (…). An entity shall classify other liabilities as non-current".

As stated before (Sect. 2.1), although equity is defined as a residual, it may be sub-classified in the balance sheet into various types of capital and reserves, such as shareholders' capital, retained earnings, statutory reserves, tax reserves (Conceptual framework, §4.45) [26]. In a simple way, equity increases through investments by shareholders and positive net income (profit) from business operations, and decreases through negative net income (loss) and dividends' payment. So, the net income of the current accounting period is a component of equity.

Profit or loss statement

The profit or loss statement reports on a company's performance over a period of time (between two balance sheet dates), presenting the income, expenses and the difference between them. This difference (income less expenses) yields the bottom-line net income amount. When income exceeds expenses, the entity got positive net income (after tax) or profit during the accounting period; otherwise, the entity got a negative net income or loss. At the end of the accounting, period net income or loss amount is transferred to equity in the balance sheet.

Statement of changes in equity

The statement of changes in equity includes, for each component of equity, a reconciliation between the carrying amount at the beginning and the end of the period (IAS 1, §106) [25].

Statement of cash flows

A statement of cash flows summarizes information concerning the cash inflows (receipts) and outflows (payments) for a specific period of time, providing "users of financial statements with a basis to assess the ability of the entity to generate cash and cash equivalents and the needs of the entity to utilise those cash flows" (IAS 1, §111) [25].

Notes

In Notes, the entity explains and gives more details that complement the remaining financial statements. Notes shall present information about the basis of preparation of the financial statements and the specific accounting policies used, disclose the information required by accounting standards that is not presented elsewhere in the financial statements, and provide information that is not presented elsewhere in the financial statements, but is relevant to an understanding of any of them (IAS 1, §112) [25]. Notes should be systematically presented, and each item in the statements should be cross-referenced to the relevant note.

Links between financial statements

Financial statements reflect the economic activity of the company and are a means to communicate the companys' financial strength and the profitability of the business. Financial statements provide a comprehensive and synthetic portrait of the business and are able to satisfy the users' different needs for information. Financial statements are intricately linked, as shown in Fig. 2.

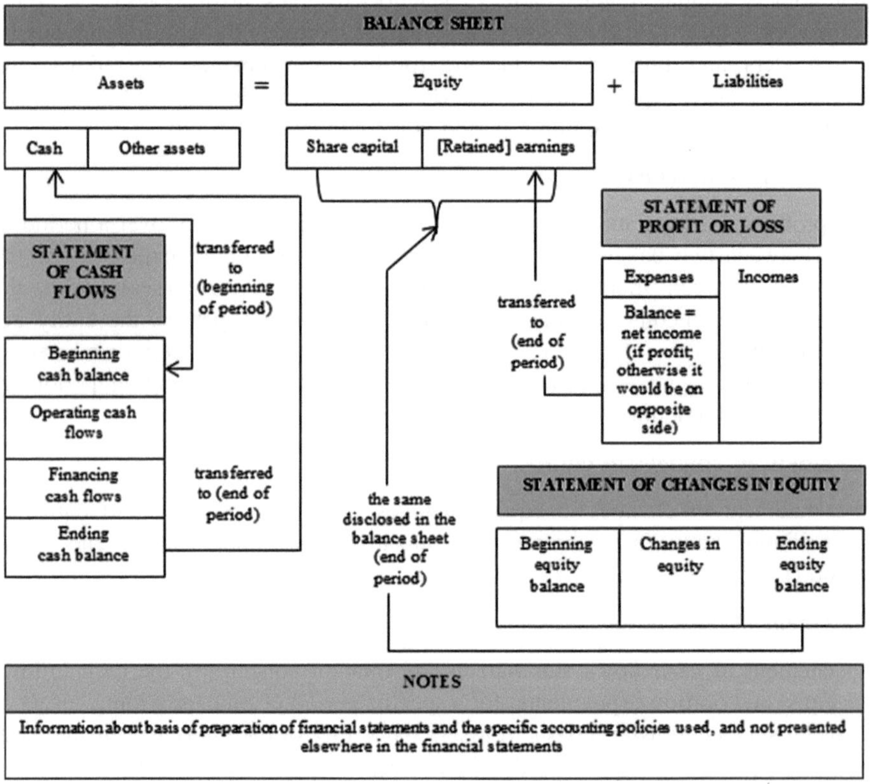

Fig. 2 Links between financial statements. *Source* [4, p. 92] adapted

3 Cost Accounting and Management Accounting

3.1 Fundamentals of Management Accounting

As mentioned before, financial accounting provides information for external users, notwithstanding accountants also seek to produce the information necessary for internal users to have a better understanding about how they can increase value to the organization (entity).

The business performance is considerably better as more appropriate and effective is the accounting information system, and the performance of accounting information system will be better when its contribution to decision-making leads to improved business performance [36]. Therefore, and different from financial accounting, as described previously, management accounting, or cost accounting, produces information that is designed for managers, and "because the managers are making decisions only for their own organization, there is no need for the information to be comparable to similar information in other organizations" [37, p. 6].

As Fig. 3 represents, the accounting processes, which aim to identify, measure and communicate all relevant economic information for the assessment and assist the decision-making, are the objects of the study of management accounting. Therefore, the performance of management will increase when the management accounting processes achieve the purposes of its existence. For this to happen,

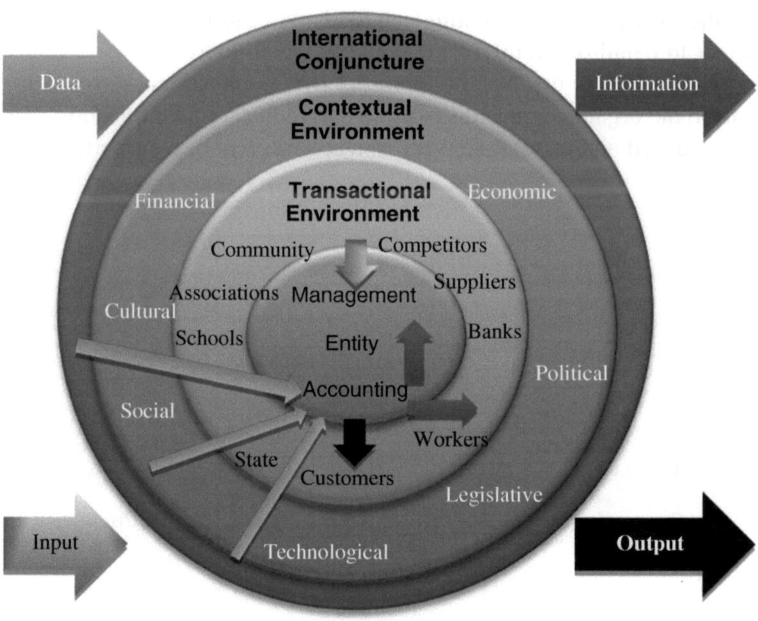

Fig. 3 Accounting information system. *Source* [36, p. 30] adapted

management accounting must seek to adapt the models, tools and techniques so that the information produced is appropriate and necessary for decision-makers.

Management accounting, also named internal accounting, aims to produce the required information to help managers developing a better understanding of an entity's costs. This is the first step for managers to add value to the organization. The information provided by management accounting is useful for all kinds of organizations, whether for small or larger organizations, profit and non-for-profit organizations, or even government agencies. However, the benefits from an improved cost/management accounting system "come from better decision making. If the benefits do not exceed the cost of implementing and maintaining the new accounting system, managers will not implement it" [37, p. 9].

Since the goal of management accounting is to assist the decision-makers (managers) in obtaining the maximization of the entity's value, internal accounting seeks to anticipate the impact of future decisions of decision-makers. In particular, management accounting pays considerable attention to cost and the way income of a specific period is obtained. However, accounting allows the establishment of three types of performance results through different perspectives: (i) economic competence; (ii) financial competence and (iii) cash competence.

In this context, and as presented in Fig. 4, the performance result obtained by the economic competence flow is obtained by subtracting costs to the income. On the other hand, the performance result for the financial competence flow is obtained by subtracting the expenses to the revenues. And yet, the performance result obtained by the cash flow competence is obtained by subtracting payments to the receipts.

Although the definitions of accounting concepts have been provided earlier based on the international accounting standards, in cost/management accounting it is important to detail differently some concepts, as follows:

Income: Income is the benefit obtained by an intentional asset delivery of activity from the organization and susceptible to monetary quantification (e.g. value of goods produced and/or sold, delivery of a service, rent of a property, exploration grant).

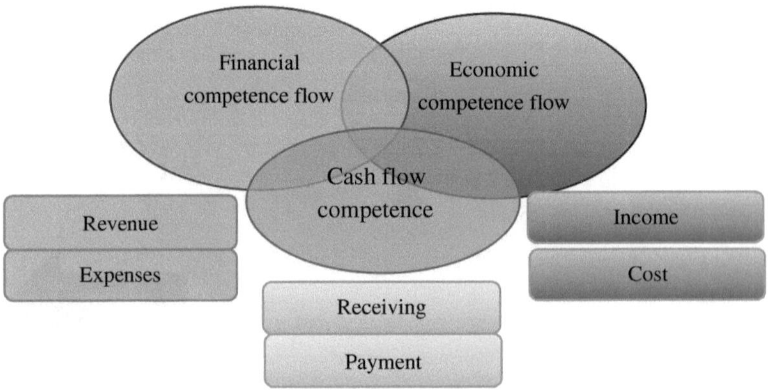

Fig. 4 Competence flow

Revenue: Revenue is the right to receive a monetary amount or equivalent (e.g. sales invoice, attributed subsidies).

Receiving: Receiving is the effective input on cash or equivalent means (implies a receipt or equivalent document).

Cost: A cost is an intentional sacrifice of resources. The cost is related to the core of the business and susceptible to monetary quantification and generally associated with the expectation to obtain a certain income (e.g. consumption of raw materials, personnel costs, depreciation). A cost can be classified as direct or as indirect.

Expenses: Expenses is the obligation (commitment) of payment (e.g. purchase invoice of raw materials, payroll processing, statement of damages, fines, income tax statement).

Payment: Payment is the effective delivery of cash or equivalent means (implies a receipt or equivalent document).

The focus of cost accounting is on costs, not expenses or payments. It is not important if we pay by cash or use another asset and whether we pay now or later. Accordingly, the income statement within cost/management accounting is prepared by adopting the economic competence flow, also named income statement by function.

3.2 Cost of Goods Manufactured and Sold

Since the financial accounting has as main objective to prepare information for outside partners, the information it produces is not sufficient for internal decision-makers, such as for the valuation of inventories or for the determination of costs by activity, by department or by sections. In this regard, within cost/management accounting, the costs are reclassified by their functions in the organization.

The components of manufactured products are the direct materials, the direct labour cost and the manufacturing overhead costs, as presented in Fig. 5. The work finished during the period is transported from the production section to the finished goods storage to be sold. On the other hand, when the products are not finished, in a specific period, its production cost is considered in the Work-in-Process inventory

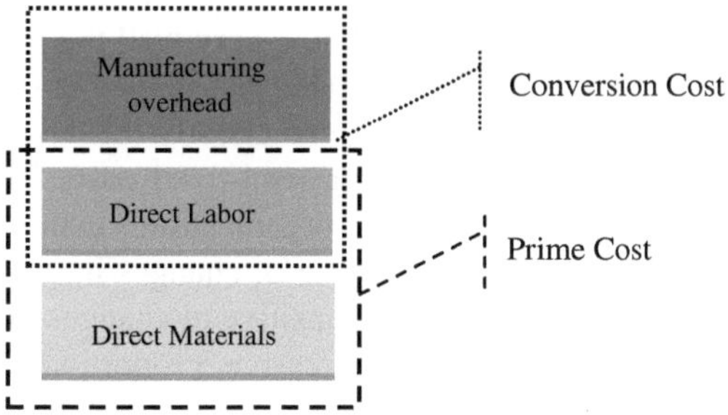

Fig. 5 Components of manufactured product cost

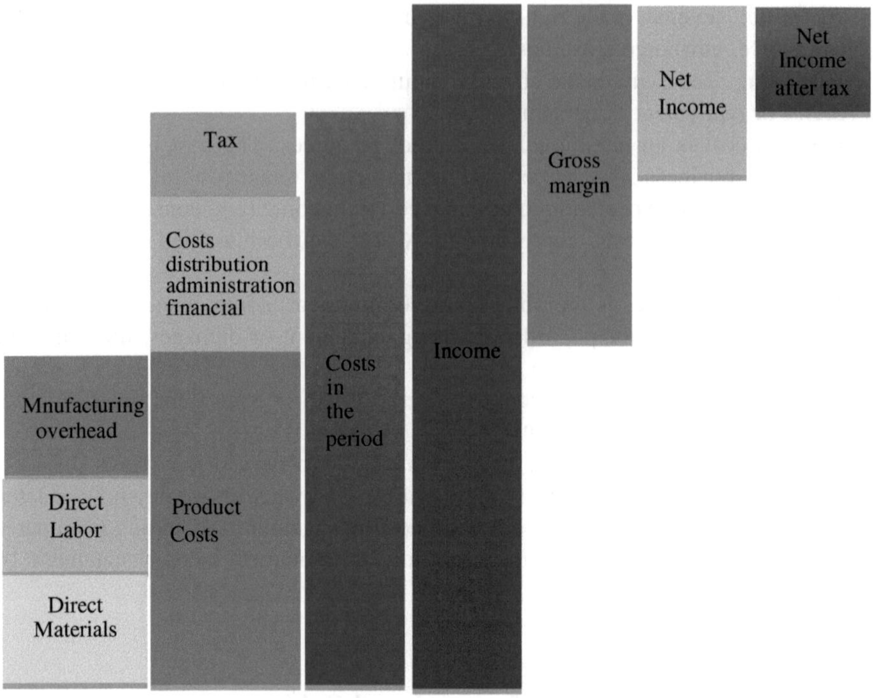

Fig. 6 Statement of the cost of goods manufactured and sold

account. Finally, the cost of goods manufactured and sold is determined as presented in Fig. 6.

3.3 Fundamentals of Cost–Volume–Profit Analysis

Cost Behaviour Patterns

Changes in volume have a significant impact on operating leverage of organizations. This sensitivity is greater when the cost structure has high fixed cost. Therefore, it is very important that managers know the cost structure of its organization. They have to know very well the potentialities or dangers that changes in the level of production/sales can induce in operating leverage [38].

In a general way, we can classify the costs as: (i) variable costs; (ii) fixed costs; and (iii) semi-variable or semi-fixed costs. This classification attends to their sensitivity to changes in the organization's activity level. Thus, fixed costs are the ones that do not change because of changes in the level of activity. In turn, the variable costs are those that are sensitive to changes in the organization's activity level, and usually they can be progressive, proportional or regressive, as presented in Fig. 7.

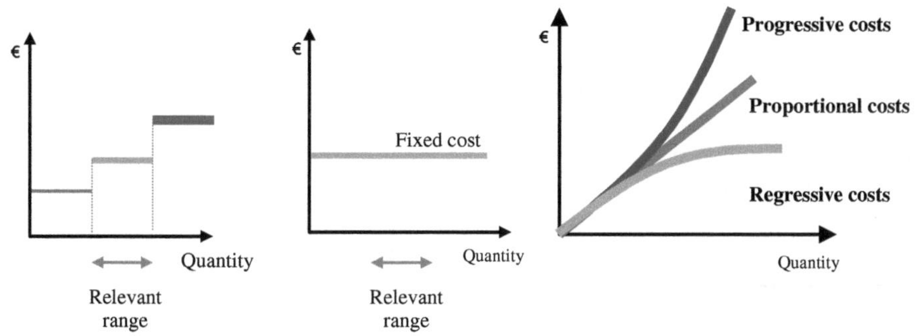

Fig. 7 Fixed and variable costs

The concept of coverage or contribution margin

The coverage or contribution margin (CM) is the excess of the value of sales after subtracting the variable costs. This margin shows the net value of individual coverage of each product. We can calculate the unit or total amount of the contribution margin. Also, we can calculate the contribution margin ratio, which represents the percentage of sales revenue net (without the variable costs).

The analysis of the contribution margin exposes how changes in the number of units sold modify the operating income. This knowledge enables the manager to decide which quantities or which goods will contribute to a better performance of the organization. Besides, managers can understand more clearly the direct consequences that a reduction/increase in the selling price will have in the reduction/increase in the operating income. Thus, this analysis will allow managers to make better decisions about the best mix of products and the best choice for the selling price.

Finding Breakeven point

The breakeven point tells us the value and quantity of sales resulting in a zero operating income. In order to determinate the breakeven point, it is necessary to check the following requirements: (i) it is possible to classify the costs as fixed or variable; (ii) variable costs are proportional to the manufacturing; (iii) fixed costs remain unchanged during the period of analysis; (iv) the unit selling price is fixed; (v) the change in manufacturing does not exist or is insignificant; (vi) the sales value is the only income; and (vii) the cost can be translated by a linear regression.

Cost–volume–profit analysis is demonstrated in Fig. 8. We can observe the behaviour of sales, variable costs and fixed costs when changing the quantities sold. The breakeven point occurs when the line of total costs crosses the line of sales. Otherwise, the breakeven point occurs when the contribution margin is enough to cover the fixed costs.

We can also determine the value and quantity of the equilibrium point using the formulas presented in Fig. 9. Moreover, it is also interesting to calculate and to analyse the **margin of safety**, because this margin represents the excess of actual (or projected) sales over the breakeven point.

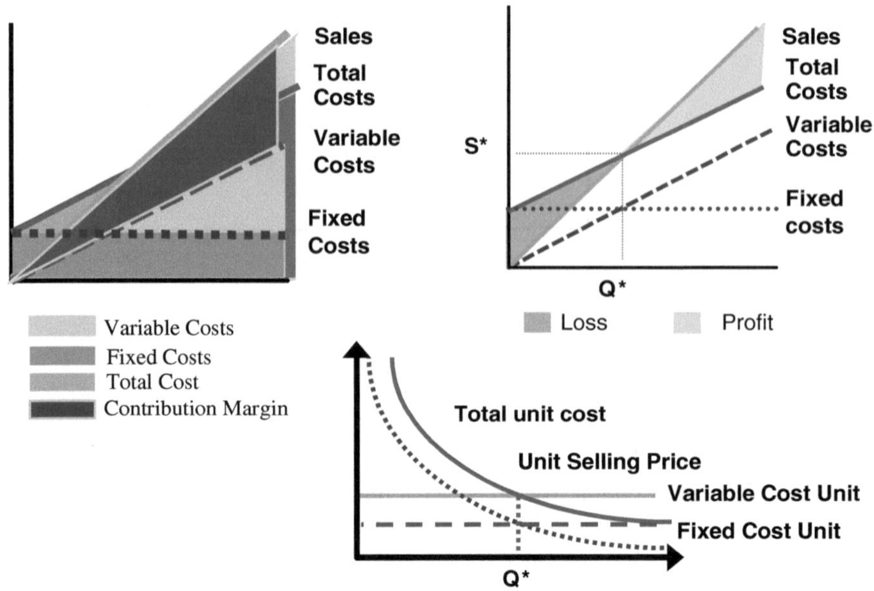

Fig. 8 Cost–volume–profit analysis

4 Review

4.1 Learning Outcomes

At the end of this chapter, you should be able to:

1. Identify the purpose of accounting.
2. Distinguish between single entry bookkeeping and double entry bookkeeping.
3. Describe the origins of double entry bookkeeping.
4. Describe the primary groups of users at which general-purpose financial statements are aimed.
5. Identify the qualitative qualities that make financial information useful.
6. Define the basic elements of financial statements—assets, liabilities, equity, income and expenses.
7. Define the criteria recognition of the financial statement elements and identify the historical cost and the current value bases as the different measurement bases of the financial statements.
8. Understand the accrual basis and going concern assumptions which underlie the preparation of financial statements.
9. Identify IFRS as the accounting and reporting standards followed in a world-wide basis.

$$\textbf{Operating income} = \text{Sales revenues} - \text{Variable cost} - \text{Fixed Cost}$$

$$\textbf{Operating income} = \text{Selling Price} \times \text{Quantity of units sold} - \text{Variable cost per unit} \times \text{Quantity of units sold} - \text{Fixed Cost}$$

$$\textbf{Contribution margin} = \text{Sales revenues} - \text{Variable cost}$$

$$\textbf{Contribution margin per unit} = \text{Selling price} - \text{Variable cost unit}$$

$$\textbf{Contribution margin percentage} = \frac{\text{Selling price} - \text{Variable cost per unit}}{\text{Selling price}}$$

Breakeven point

$$\textbf{Breakeven quantity of units} = \frac{\text{Fixed costs}}{\text{Contribution margin per unit}}$$

$$\textbf{Breakeven quantity of units} = \frac{\text{Fixed costs}}{\text{Selling price} - \text{Variable cost unit}}$$

$$\textbf{Breakeven sales revenues} = \frac{\text{Fixed costs}}{\text{Contribution margin percentage}}$$

$$\textbf{Breakeven sales revenues} = \text{Breakeven quantity of units} \times \text{Selling price}$$

Margin of safety

$$\textbf{Margin of safety} = \text{Sales actual (or projected)} - \text{Breakeven sales}$$

$$\textbf{Margin of safety percentage} = \frac{\text{Sales actual (or projected)} - \text{Breakeven sales}}{\text{Sales actual (or projected)}}$$

Fig. 9 Formulas for cost–volume–profit analysis

10. Identify and describe the components of a complete set of financial statements. Understand the links between the financial statements.
11. Distinguish between financial accounting and management accounting.
12. Identify the usefulness and users of cost/management accounting information.

5 Multiple-Choice Questions

1. The objective of financial reporting places most emphasis on:

   (a) reporting to capital providers.
   (b) reporting to tax authorities.
   (c) providing specific guidance related to specific needs.
   (d) providing information to individuals who are experts in the field.

2. General-purpose financial statements are prepared primarily for:
   (a) internal users.
   (b) external users.
   (c) tax authorities.
   (d) auditors.
   (e) government regulators.
3. Net income will result during a time period when:
   (a) assets exceed liabilities.
   (b) assets exceed revenues.
   (c) expenses exceed income.
   (d) income exceed expenses.
4. Identify the qualitative characteristic(s) that relate(s) to the following sentences:
   (a) Two fundamental qualities that make accounting information useful for decision-making purposes.
   (b) Qualitative characteristic being displayed when companies in the same industry are using the same accounting principles.
   (c) Ignores the economic consequences of a standard or rule.
   (d) Imperative for providing comparisons of a company from period to period.
   (e) Predictive value is an ingredient of this fundamental quality of information.
   (f) Requires a high degree of consensus among individuals on a given measurement.
   (g) Neutrality is a key ingredient of this fundamental quality of accounting information.
5. Identify if the following sentences are true or false. Justify.
   (a) The steps in the accounting process are identification, recording, control and communication.
   (b) The cost principle dictates that companies record assets at their cost and maintain that until its derecognition.
   (c) Fair value measurement must be used if fair value is higher than its cost.
   (d) The balance sheet is prepared as of a specific date.
   (e) Expenses are recognized when incurred, rather than when paid.
   (f) Revenues are recognized when cash is received.
   (g) Current assets are the ones' that a company expects to convert to cash or use up within one year or the operating cycle, whichever is longer.
   (h) The statement of financial position is the primary source for evaluating a company's performance.
   (i) When a company earns incomes, equity increases.

6. Double entry bookkeeping emerged:

   (a) In the twentieth century.
   (b) To meet the needs with the growing business requirements around the fourteenth century.
   (c) As an invention made by Fra Luca Pacioli.
   (d) But was abandoned for its complexity.

7. Accounting has a long history because:

   (a) Its origins are placed in the nineteenth century with the industrial revolution.
   (b) Its origins are placed in the Middle Age.
   (c) Its origins are placed in ancient civilizations, dating back to about 4000–3000 B.C.
   (d) Its origins are commented with the printing press.

8. Identify if the following sentences are true or false. Justify.

   (a) Financial accounting information provides information to users (decision-makers) who are involved in the operations and strategy of the firm. These users are often internal to the firm.
   (b) Although cost accounting information is often used in the financial accounting system, its primary role is to aid managers inside the firm in making operational and strategic decisions.

9. Identify if the following sentences are true or false. Justify.

   (a) A cost is something used up to produce revenues in a particular accounting period.
   (b) Variable costs are those that are not sensitive to changes in the organization's activity level.

10. Identify if the following sentences are true or false. Justify.

   (a) The breakeven point tells us the value and quantity of sales resulting in a zero operating income, information with considerable relevance to decision-makers.
   (b) The requirements to calculate the breakeven point are so simplistic that from its determination it is not possible to make any analysis.

Multiple-Choice Answers

1. a;
2. b;
3. d;

4a. Relevance and faithful representation;
4b. Comparability;

4c. Neutrality;
4d. Comparability;
4e. Relevance;
4f. Verifiability;
4g. Faithful representation;
5a. True;
5b. False;
5c. False;
5d. True;
5e. True;
5f. False;
5g. True;
5h. False;
5i. True;
6. b;
7. c;
8a. False;
8b. True;
9a. True;
9b. False;
10a. True;
10b. False.

References

1. Gomes D (2008) The interplay of conceptions of accounting and schools of thought in accounting history. Acc Hist 13(4):479–509
2. Meyer JW (1986) Social environments and organizational accounting. Acc Organ Soc 11 (4/5):345–356
3. Riahi-Belkaoui A (2007) Accounting theory, 5th edn. Thomson Learning, London
4. Stolowy H, Lebas M, Ding Y (2013) Financial accounting and reporting—a global perspective, 4th edn. Seng Lee Press, Singapore
5. Hopwood AG (1983) On trying to study accounting in the contexts in which it operates. Acc Organ Soc 8(2/3):287–305
6. Miller P (1994) Accounting as a social and institutional practice: an introduction. In: Hopwood AG, Miller P (eds) Accounting as social and institutional practice. University Press, Cambridge, pp 1–39
7. Potter BN (2005) Accounting as a social and institutional practice: perspectives to enrich our understanding of accounting change. Abacus 41(3):265–289
8. American Accounting Association (AAA) (1966) Committee to prepare a statement of basic accounting theory. Statement of basic accounting theory. AAA, Sarasota, Fl
9. Yamey BS (1980) Early views on the origins and development of book-keeping and accounting. Acc Bus Res 10:81–92
10. Littleton AC (1933) Accounting evolution to 1900. American Institute Publishing Co., Inc, New York
11. Edwards JR (1989) A history of financial accounting. Routledge, London

12. Vangermeersch R (1996a) Single entry bookkeeping. In: Chatfield M, Vangermeersch R (eds) The history of accounting, an international encyclopedia. Garland Publishing Inc, New York, London, p 533
13. Poppof B (1996b) Income—determination theory. In: Chatfield M, Vangermeersch R (eds) The history of accounting, an international encyclopedia. Garland Publishing Inc, New York, London, p 311
14. Vangermeersch R (1996) Ledger. In: Chatfield M, Vangermeersch R (eds) The history of accounting, an international encyclopedia. Garland Publishing Inc, New York, London, pp 372–373
15. de Roover R (1956) The development of accounting prior to Luca Pacioli according to the account-books of medieval merchants. In: Littleton AC, Yamey BS (eds) Studies in the history of accounting. Sweet & Maxwell, London, pp 114–174
16. Hernández Esteve E (1992) Propuestas de contabilidad por partida doble para llevar las cuentas centrales de la Real Hacienda Castellana (Hacia 1574). Técnica Contable XLIV (535–554):649–664
17. Mills GT (1994) Early accounting in Northern Italy: the role of commercial development and the printing press in the expansion of double-entry from Genoa, Florence and Venice. Acc Historians J 21(1):81–98
18. Lee GA [1977] (1984) The coming age of double entry: the Giovanni Farolfi Ledger of 1299–1300. Acc Historians J, Fall, pp 79–95 (Reprinted in Nobes C (ed) The development of double entry bookkeeping, selected essays. Garland Publishing Inc, New York, London, pp 45–95)
19. Chatfield M (1996) Massari commune ledgers. In: Chatfield M, Vangermeersch R (eds) The history of accounting, an international encyclopedia. Garland Publishing Inc, New York, London, p 400
20. Yamey BS (1940) The functional development of double-entry bookkeeping. The Accountant, pp 333–342
21. Amorim JL (1968) Digressão através do vetusto mundo da contabilidade. Livraria Avis, Porto
22. Gomes D (2007) Accounting change in central Government: the institutionalization of double entry bookkeeping at the Portuguese Royal Treasury (1750–1777). Ph.D. dissertation, University of Minho, Braga, Portugal. Available at: https://repositorium.sdum.uminho.pt/bitstream/1822/6754/1/Tese%20Doutoramento%20Delfina%20Gomes.pdf
23. Chatfield M (1977) A history of accounting thought. Robert E. Krieger Publishing Company, New York
24. IASB (1989) Conceptual framework for financial reporting, London
25. IASB (2014) International Accounting Standard No. 1 (IAS 1) Presentation of Financial Statements, London
26. IASB (2015a) Exposure Draft ED/2015/3 Conceptual Framework for Financial Reporting. http://www.ifrs.org/Current-Projects/IASB-Projects/Conceptual-Framework/Documents/May%202015/ED_CF_MAY%202015.pdf. Accessed 7 Aug 2015
27. Barth M (2007) Standard-setting measurement issues and the relevance of research. Acc Bus Res Spec Issue Int Acc Policy Forum 37(1):7–15
28. Laux C, Leuz C (2009) The crisis of fair value accounting; making sense of the recent debate. Acc Organ Soc 34(6/7):826–834
29. IASB (2011) International Financial Reporting Standard No. 13 (IFRS 13) Fair Value Measurement, London
30. Parker R (2012) Introduction. In: Nobes C, Parker R (eds) Comparative international accounting. 12th edn. Prentice Hall, London, pp 3–26
31. Radebaugh L, Gray S, Black E (2006) International accounting and multinational enterprises, 6th edn. Wiley, New York
32. IASB (2015b) Who we are and what we do. http://www.ifrs.org/About-us/Pages/Who-We-Are.aspx. Accessed 30 July 2015

33. European Commission (EC) (2002) Regulation No. 1606/2002 of the European Parliament and of the Council of 19 July 2002 on the application of International Accounting Standards. Official Journal of the European Communities, L243 11/09/200, pp 0001–0004, Brussels, Belgium. http://eur-lex.europa.eu/legal-content/EN/ALL/?uri=CELEX:32002R1606. Accessed 4 Aug 2015
34. Nobes C (2006) The survival of international differences under IFRS: towards a research agenda. Acc Bus Res 36:233–245
35. Nobes C (2013) The continued survival of international differences under IFRS. Acc Bus Res 43(2):83–111
36. Silva AF, Silva AM (2005) Áreas disciplinares da Contabilidade, *Revista TOC*, 69, December, pp 29–36
37. Lanen WN, Anderson SW, Maher MW (2011) Fundamentals of cost accounting, 3rd edn. McGraw-Hill/Irwin, New York
38. Lanen WN, Anderson SW, Maher MW, Dearman DT (2010) Fundamentals of cost accounting. Issues Acc Educ 25(4):791–792

Introduction to Corporate Finance

Leoni Eleni Oikonomikou

Abstract This chapter provides a comprehensive overview of the main topics discussed in a first-year MBA class on corporate finance. One more motivation is not only to summarize them but also to provide some hands-on tools that can be applied in practice in relevant to corporate finance positions. Section 1 introduces the reader to the corporate finance field, the organizational structure of the firm, the conflicts between stakeholders, the financial management decisions as well as the financial markets. Section 2 refers to the utility of financial statements and provides the way to extract the necessary information from them. Later on, we discuss about the cost of capital and its fundamental importance in the capital structure of the firm. Finally, we present some tools for capital budgeting decisions and project valuation such as net present value, internal rate of return, and discounted payback period. Attention is also paid to the dividend policy of the firm and its consequences to the firm's earnings. The chapter concludes by presenting some of the most important techniques in equity and business valuation such as free cash flow and discounted cash flow analyses.

1 Introduction

Corporate finance is a specialization in finance that deals with the sources of funding, the capital structure of a corporation/entity, the actions that managers take to increase the shareholder value of the firm, as well as the tools and analysis used to allocate financial resources (Wikipedia).

In academic terms, the principal goal of a Corporation is to maximize shareholder value. It is also referred as value maximization or Net Present Worth maximization. This goal concerns the amount and share of national income which is

L.E. Oikonomikou (✉)
Georg-August-Universität Göttingen, Platz der Göttingen Sieben 3,
37073 Göttingen, Germany
e-mail: leoni-eleni.oikonomikou@wiwi.uni-goettingen.de; le.oikonom@gmail.com

C. Machado and J.P. Davim (eds.), *MBA*,
Management and Industrial Engineering, DOI 10.1007/978-3-319-28281-7_6

paid to the owners of the firm. Therefore, managers should take the appropriate decisions that maximize the shareholder wealth or generate a Net Present Value (NPV).

Shareholder value is a situation where the value created by the use of resources is more than the total of input resources. Moreover, the profit maximization goal orients the investment, financing, and dividend decisions of the firm toward that. However, profit maximization is a vague and ambiguous concept. It ignores the differences in terms of the timing of benefits acquired from investment proposals. And it also ignores the quality of benefits associated with an investment proposal.

There are two approaches in corporate finance: the *traditional* and the *modern* one. The *Traditional approach* defines corporate finance as the management of financial resources of a corporation/business entity. Alternatively, it is also referred as *collection of funds approach* since it limits the finance functions only to the procurement funds and ignores the use of the other types of funds.

Corporate finance, within its *Modern Approach*, does not only deal with financing decisions, but also with investment and current management decisions. It is considered as an understandable and internationally accepted approach with the procurement of funds and its efficient usage.

Therefore, corporate finance is about planning, raising, investing, and monitoring the finance in a corporation.

1.1 Organizational Structure of the Firm

Corporate finance applies to all types of businesses. The financial management of a corporation/entity differs according to its type: *Sole Proprietorship*, *Limited Liability Company*, *Partnership*, *Trust*, *Corporation*, *Small and Medium* (SME) [7].

The decisions are taken by the owners in SMEs and by professional managers in large companies. Therefore, the authority/responsibility in running the financial management of a corporation/entity belongs to the *chief financial officer* (CFO). The *CFO or financial manager* is the buyer of capital (via the traditional corporate finance approach). He/She negotiates with various investors (bankers, shareholders, bond investors) in order to obtain the necessary funds to run the business at the lowest attainable cost. These transactions usually take place in the capital markets.

The *financial manager* also establishes the appropriate relationship among people involved in finance functions within the organization, assures division of work, and prevents confusions on roles and responsibilities of employees, duplication, and overlapping of activities. The *treasurer* follows in hierarchy and oversees *financial planning*, *capital expenditures*, *cash management*, and *credit management* within a firm. Finally, the *Controller* monitors *financial, cost accounting*, *and taxation*.

Figure 1 is taken by the Webpage www.slideshare.net and shows a business organ gram. Its practical usage is well acknowledged within practitioners as a way to avoid role confusions.

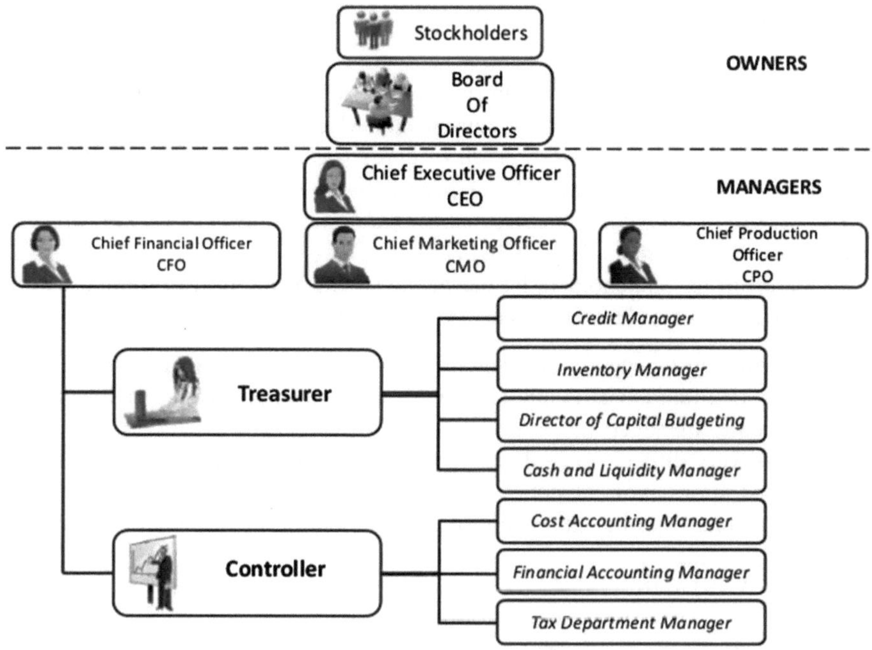

Fig. 1 Organ gram of a business entity (in this chapter, the focus is on the *left side* that represents the financial management of the firm and its managers)

The *financial management division* within a corporation accounts for the following functions:

I. planning and budgeting,
II. allocating resources,
III. operating, monitoring and safeguarding, and
IV. evaluating and reporting.

From a practitioner's point of view, the possible goals of *Financial Management* are various. First of all, it includes the survival of the firm and the fight against the competition. It also targets on the maximization of sales, net income, market share, and the value of investment shares while it also aims at minimizing costs.

From an academic point of view, the goal of *financial management* is to maximize the fundamental or economic value of investment shares. This goal stems from the fact that shareholders own shares and the principal goal of the firm is to maximize shareholder value. Manager's goal is to act in such a way as to accomplish this principal goal, as well. However, the limitation comes from the fact that this value is not directly observable.

Some academics consider that the principal goal of a corporation should be stock price maximization (as previously mentioned) rather than wealth maximization. Therefore, these two are deemed as equivalent.

1.2 Agency Problem

The *principal* (shareholders)—*agent* (manager) *problem*, *agency theory*, or *agency dilemma*—was first introduced by Jensen and Meckling, in 1976. According to Wikipedia, "*it occurs when one person or entity (agent) is able to make decisions on behalf of, or that impact, another person or entity (principal). The dilemma exists because the agent is sometimes motivated to act in his/her own best interests rather than those of the principal.*" [19]

In lots of firms, shareholders let the management of the firm to foreign control (managers/directors). However, based on this theory, this leads to a risk:

- in determining managerial accountability when letting managers run the firm.
- Another risk that exists is that shareholders are at information disadvantage as compared to managers.
- It is also very difficult to evaluate how well a manager has performed since he possesses information advantage over the shareholders.
- Finally, it takes significant amount of time to monitor the results of the decisions managers made.

1.3 Conflicts Between Company Stakeholders

As previously mentioned, in theory, managers should work, in the best interest of the shareholders. Unfortunately, this does not always happen in practice, since managers may maximize their own wealth (in the form of high salaries) at the cost of shareholders, buy other companies to expand their power, venturing onto fraud, manipulate financial figures to comment on bonuses and stock price-related options.

There are various solutions in order to resolve the conflicts between the shareholders and the managers:

- First, comes the increase of managerial compensation.
- Second, shareholders may intervene directly on to the matter.
- Third, there is a threat of firing.
- Fourth, there is a threat of hostile takeovers [4].

These conflicts do exist between shareholders and creditors as well. Shareholders via managers make decisions by ignoring the interest of creditors. Therefore, a manager may decide to invest in a risky project. If this project succeeds, all benefits go to its shareholders and creditors will only receive the already fixed low rate of return. On the contrary, if this project fails, creditors have to share the losses with all the other parties.

In order to resolve these conflicts between shareholders and creditors, creditors may request an additional compensation for the increased risk they are bearing or they may request protective terms and conditions into the contract they are signing with shareholders.

1.4 Financial Management Decisions

The three types of *financial management decisions* that corporate finance seeks to answer are the following:

- *Capital budgeting*,
- *Capital structure*, and
- *Working capital management*.

Within the Capital budgeting decisions, the firm seeks to answer:

- What long-term investments or projects should a firm take on?

Capital budgeting is a process of planning and managing the long-term investments of a firm [3]. In such type of decisions, a financial manager identifies potential investment opportunities that are profitable for the firm. For instance, a technology firm that decides on whether or not to open a new branch is a capital budgeting decision. Some of the main issues that capital budgeting deals with are the size of cash inflows and outflows of such a potential project, timing of cash flows and riskiness of them.

According to the capital structure decisions:

- How should a firm pay for its assets?
- Is it better to use debt or equity?

Capital structure refers to the combination of debt and equity a company uses to finance its long-term operations and growth. The proportions of short- and long-term debt are taken into account in the capital structure analysis of an entity [14]. The most insightful financial ratio in this case is the entity's debt-to-equity ratio that provides information about the level of riskiness of this corporation. It is common that an entity heavily financed by debt is perceived as riskier, due to its high leverage levels [15]. However, at the same time, this entity might enjoy significant tax benefits [1].

Finally, based on the working capital management decisions:

- How would the firm manage its everyday financial activities?

Working capital management refers to the firm's short-term assets including inventory and liabilities. It is a day-to-day operation. It involves planning and managing the firm's current assets and liabilities.

Some examples of working capital management include:

- How should the firm sell to on credit?
- How much inventory should it carry?
- When should it pay its suppliers?

Indeed, financial managers seek to answer some or all of the following questions: How should firms allocate scarce resources to minimize expenses and maximize revenues? How should they finance their activities—via stocks or bonds, equity (owner capital), or debt (bank loan)? How should a firm use its profits? How much of it should it be reinvested into the company, and how much should it pay out to its shareholders?

1.5 Financial Markets

Financial Markets refer to an exchange where firms refinance their activities.

Cash enters the firm via the sale of debt or equity. This money is used to purchase assets. These assets generate cash that is used to pay stakeholders, reinvest in additional assets, repay debtholders, and pay dividends to stockholders/equityholders [23].

Three main types of exchange markets exist: *primary market, secondary market,* and *over-the-counter market.*

Primary market is where a newly issued security is first offered. Companies, governments, and other groups obtain financing through debt- or equity-based securities.

A *Secondary Market* is the stock market where investors purchase securities or assets from other investors, rather than from issuing companies themselves.

According to NASDAQ, an *Over-the-counter market* is a decentralized market where geographically dispersed dealers are linked by telephones and computers. The market trades securities not listed on a stock or derivatives exchange.

Firms issue stocks and bonds:

A *stock/equity* is a type of security that demonstrates ownership in a corporation and represents a claim on part of the corporation assets and earnings. Stocks are categorized as *common* and *preferred stocks.*

Common stock/equity gives the owner the right to vote at shareholders' meetings and receive dividends. *Preferred stock/equity* has no voting rights but a higher claim on assets and earnings than *common shares*. Holders of preferred stock/equity receive dividends before common stockholders and have priority in case of company's default and liquidation.

A *bond* is a debt instrument issued by governments, corporations, and other entities in order to finance projects or activities. Bonds are commonly referred to as *fixed-income securities.* Many corporate and government bonds are publicly traded on exchange markets, while others are traded only *over-the-counter* (**OTC**) [12].

1.5.1 Financial Market Efficiency

One of the most well-known topics in Financial Market Theory is the *Efficient market hypothesis* (**EMH**). This theory was developed by Eugene Fama in 1970s. It stands as of crucial importance for stock market investors who are usually wondering whether the market is efficient, i.e., whether it reflects all the information made available to market participants at any given point in time. Based on the

EMH, all stocks are perfectly priced according to their inherent investment properties, the knowledge of which all market participants possess equally. However, as every theory, EMH has also its deficiencies which are enumerated below.

- Firstly, EMH assumes that all investors perceive all available information in almost the same way. One main reason for that may be the numerous methods used to analyze and value stocks and the problems they impose on the validity of the EMH.
- Secondly, under the EMH, no single investor is able to attain greater profitability than another with the same amount of investments. This equal possession of information is interpreted as that investors can only achieve identical returns.
- Thirdly, under EMH, no investor should ever be able to beat the market.

All the above imply that markets are not truly efficient. For greater efficiency to occur, the following prerequisites must be met:

- universal access to high-speed and advanced systems of pricing analysis,
- universally accepted analysis system of pricing stocks,
- absolute absence of human emotion in investment decision making,
- investors' willingness to accept that their returns or losses will be exactly identical to all other market participants.

1.5.2 Fundamental Concepts of Risk and Return

– *Arithmetic mean return*

Arithmetic mean return (**AM**) is usually referred to as a simple average or mean and is calculated as such

$$\mathrm{AM} = \frac{\sum_{i=1}^{n} R_t}{T} \tag{1}$$

where

$\mathbf{R_t}$ simple return in period **t** and

T the number of returns.

– *Geometric mean return*

Geometric mean return (**GM**) is referred to as the compound *annual growth rate* or *time-weighted rate of return*, as well. It is the average rate of return on a given set of values calculated using the products of the terms. Its difference from the previous AM is that GM takes into account the effects of compounding, whereas AM fails to do that. In addition, AM is always higher than GM in value. It is given by the following formula:

$$\mathrm{GM} = \prod_{i=1}^{T} (1 + R_i)^{\frac{1}{T}} - 1 \tag{2}$$

– *Standard deviation*

Standard deviation (SD) is considered as a measure of total risk. It measures the dispersion of a set of returns from its mean.

The part of the risk in a portfolio that is diversified is called *non-systematic risk* or *idiosyncratic risk* and consists of all those company-specific and industry-specific factors that affect the stock price of companies. The risk in a portfolio that cannot be diversified consists of changes in interest rates, macroeconomic factors, and political events. The *standard deviation* can be calculated by the following formula:

$$\mathrm{SD} = \frac{\left[\sum_{t=1}^{T} (R_t - \mathrm{AM})^2\right]^{\frac{1}{2}}}{T} \tag{3}$$

where

R_i return in period **T**
AM arithmetic mean return of the asset over the period considered
T number of returns during that period

– *Beta*

Beta is referred to *systematic or market risk* and is measured by a stock's beta (**β**). It is considered as a relevant measure of asset's risk only when you hold the asset within a widely diversified portfolio. In that case, the asset's beta is the contribution of the asset to the volatility of the whole portfolio.

Beta is also interpreted as the reaction of stock to fluctuations in the market. It is the average return of the stock given a 1 % fluctuation in the market. Therefore, beta of 1 indicates average risk or return fluctuations similar to those of the market; a beta higher than 1 indicates higher risk (larger fluctuations) than the market and beta lower than 1 indicates lower risk (smaller return fluctuations) than the market.

In addition, based on financial market professionals, corporations' betas are usually calculated every 5 years.

– *Diversification*

The concept of diversification was first introduced by Markowitz, while developing his *capital asset pricing model* and is about combining multiple assets in a portfolio. This diversification strategy aims to protect the investor via mitigating corporate risks such as company-specific or industry-specific risks. In general, it aims to reduce non-systematic or diversifiable risk.

– *Correlation coefficient*

The correlation coefficient is the magnitude that measures the sign and strength between two variables. When these two variables tend to move together, the correlation coefficient is positive; otherwise, it is negative. The correlation coefficient can take any value between −1 and 1 [10, 11].

2 Analysis of Financial Statements

Financial statements refer to a collection of reports regarding the entity's financial results, financial condition, and cash flows (CFs). They provide information to owners and creditors about the firm's financial performance. They can further be used by owners and creditors to set performance targets, impose restrictions on the managers of the firm, and set up a convenient template for financial planning.

- They determine the ability of a corporation to generate cash as well as the sources and uses of that cash.
- They also determine whether the company has the capacity to pay back its debts.
- They track financial results on a trend line to spot any looming profitability issues.
- Moreover, it is possible to derive financial ratios from the statements in order to determine the business condition and current performance.
- Finally, they investigate the details of certain business transactions, as outlined in the disclosures that accompany the statements.

The main contents of a set of financial statements are the following:

- *Balance sheet* shows the entity's assets, liabilities, and stockholder's equity as of the issue date.
- *Income statement* shows the results of the entity's operations and financial activities for the reporting period. It includes *revenues*, *expenses*, *gains,* and *losses*.
- *Statement of cash flows* shows changes in the entity's cash flows during the reporting period.
- *Supplementary notes* include explanations of various activities, additional detail on some accounts and items as mandated by the applicable accounting framework such as GAAP or IFRS.

At the most minimal level, a business is expected to issue a *balance sheet* and an *income statement* to document its monthly results and ending financial conditions. The full set of *financial statements* is expected when a business is reporting the results for a full fiscal year, or when a publicly held business is reporting the results of its fiscal quarters.

2.1 Balance Sheet

Balance sheet is considered as an "*inventory*" of what the firm owns (*assets*) and how those assets are financed (*liabilities and shareholder's/owner's equity*). Therefore, the following equation needs to be fulfilled:

$$\text{Assets} = \text{Liabilities} + \text{Shareholder's Equity} \tag{4}$$

Assets are resources controlled by the entity as a result of past events and from which future economic benefits are expected to flow to the entity.

Liabilities concern the list of sources where the assets came from as well as the monetary amounts of the sources from which the entity obtained its present stock of resources.

Equity, in the accounting context, means the owner's stake in the entity.

2.2 Income Statement

The *Income statement* is also known as *profit and loss statement.* It summarizes the entity's profitability over a given period of time. It records *sales*, *expenses*, *taxes,* and *net income.*

$$\text{Revenue} - \text{Expenses} = \text{Income} \tag{5}$$

The *expenses* are categorized as such:

- *Operating expenses* provide benefits only for the current period (*cost of labor and materials, depreciation based on historical cost, R&D*);
- *Financing expenses* arise from *non-equity financing* (interest expenses);
- *Capital expenses* generate benefits over multiple periods written off as depreciation (buying land and buildings).

The *total operating revenues* subtract from the sold cost of goods, selling, general, and administrative expenses, and depreciation. The operating income adds to the other income. The EBIT (earnings before interest and taxes) subtracts interest expense. The taxable income includes current and deferred taxes.

Retained earnings are not added to the cash balance in the *balance sheet* but are added to *shareholder's equity*. Therefore, net income is calculated by

$$\text{Net Income} = \text{Retained Earnings} + \text{Dividends} \tag{6}$$

However, inflation distorts the measuring of income and the valuation of assets.

2.3 Cash Flow Statement

The *Cash flow statement* reports how much cash is generated during a period and indicates where the cash comes from as well as what the firm does with that cash. Unlike the *balance sheet* and *income statement*, *cash flow statements* are independent of accounting rules as the latter have a second-order effect in cash flows via taxes.

2.4 Financial Ratios

Financial ratios help us extract the necessary information from an entity's set of *financial statements*. There are also various other methods in order to analyze financial statements such as *trend analysis* or *cross-sectional analysis*.

– ***Profitability ratios***

$$\text{Return on Assets (ROA)} = \frac{\text{EBIT}(1 - \text{tax})}{\text{Total Assets}} \quad (7)$$

$$\text{Return on Equity(ROE)} = \frac{\text{Net Income}}{\text{Book Value (Equity)}} \quad (8)$$

$$\text{Gross Profit Margin} = \frac{\text{Revenue} - \text{Cost of Goods Sold}}{\text{Revenue}} \quad (9)$$

$$\text{Operating Profit Margin} = \frac{\text{EBIT}}{\text{Sales}} \quad (10)$$

$$\text{Net Profit Margin} = \frac{\text{Net Income}}{\text{Sales}} \quad (11)$$

– ***Activity ratios*** measure the efficiency of *working capital management*:

$$\text{Accounts Receivable turnover} = \frac{\text{Sales}}{\text{Average Accounts Receivable}} \quad (12)$$

$$\text{Inventor Turnover} = \frac{\text{Cost of Goods Sold}}{\text{Average Inventory}} \quad (13)$$

$$\text{Total Asset Turnover} = \frac{\text{Sales}}{\text{Total Assets}} \quad (14)$$

- *Liquidity* measures how fast and/or easy it is to generate cash from an asset. ***Liquidity ratios*** measure short-term liquidity:

$$\text{Current ratio} = \frac{\text{Current Assets}}{\text{Current Liabilities}} \tag{15}$$

$$\text{Quick ratio} = \frac{\text{cash} + \text{short-term investment} + \text{receivables}}{\text{current liabilities}} \tag{16}$$

$$\text{Cash ratio} = \frac{\text{cash} + \text{receivables}}{\text{current liabilities}} \tag{17}$$

- ***Financial Leverage ratios***

$$\text{Debt-to-Capital ratio} = \frac{\text{Debt}}{\text{Debt} + \text{Equity}} \tag{18}$$

$$\text{Debt-to-Equity ratio} = \frac{\text{Debt}}{\text{Equity}} \tag{19}$$

The debt-to-equity ratio can be based on either *book value* (BV) or *market value* (MV). Similar condition holds for *long-term debt ratios*.

- ***Market value ratios***

The *price-to-earnings ratio* expresses the stock market price (PS) to earnings per share (EPS):

$$\text{Price-to-Earnings ratio} = \frac{\text{PS}}{\text{EPS}} \tag{20}$$

The *Dividend Yield* expresses the *latest dividend* (D) to *current stock price* (PS):

$$\text{Dividend Yield} = \frac{D}{\text{PS}} \tag{21}$$

$$\text{Market-to-Book Value} = \frac{\text{MV}}{\text{BV}} \tag{22}$$

$$\text{Market-to-Book Equity} = \frac{\text{Market Equity (ME)}}{\text{Book Equity (BE)}} \tag{23}$$

$$\text{Tobin's } q = \frac{\text{MV}}{\text{Replacement Value}} \tag{24}$$

– ***Links between the ratios***

$$\text{ROA} = \text{Profit Margin} * \text{Asset Turnover} \quad (25)$$

This ratio holds for both net and gross ROA and profit margin. An increasing ROA can be interpreted as a trade-off between profit margin and asset turnover.

$$\text{ROE} = \text{ROA} * \text{Equity Multiplier} \quad (26)$$

where

$$\text{Equity Multiplier} = \frac{\text{Assets}}{\text{Equity}} \quad (27)$$

A high *financial leverage* in an entity magnifies ROE when (gross) ROA excesses the interest on debt.

– ***Non-financial measures of operating effectiveness***

Non-financial measures can account for the following:

- innovation,
- customer service,
- product quality,
- reputation, and
- good employee relations.

3 Time Value of Money

The time value of money provides an introduction to the tools of valuation of cash flows of a project of a firm or the firm by itself. *Time value of money* (**TVM**) *principle* applies when the value of a given amount of money to be received on a specific date is more than the same amount of money to be received on a later date. In simple terms, it means that cash in the future is worth less than cash today. One reason is that money can be reinvested and generate even more money. Another reason is the inflation which is the rise in the general level in prices and results in a deterioration of our purchasing power [6].

The concepts of *discounting* and *compounding cash flows* are largely used in corporate finance. *Discounting cash flows* moves corporation's cash flows back in time while *compounding* moves them forward in time.

Concepts such as *present value* and *future value* of a single sum of money; *present value* and *future value of an annuity;* and *present value of a perpetuity* are largely used in investment valuation as well.

Present value refers to the current value of a future payment or even a series of payments that are discounted at a given interest rate up to today, in order to reflect the TVM. It is calculated based on the following formula:

$$\text{Present Value of a single sum of Money: } \text{PV} = \frac{\text{FV}}{(1+R)^T} \tag{28}$$

where

FV Future value of a single sum of money

R rate of return offered in the capital markets, alternatively referred as *discount rate*

T number of compounding periods

The *future value* reflects the value of money or cash at a prespecified future date that is also equivalent to the value of sum today. It is calculated based on the following formula:

$$\text{Future Value of a single sum of Money: } \text{FV} = \text{PV} * (1+R)^T \tag{29}$$

An ***annuity*** is a series of evenly spaced equal payments made for a certain amount of time. The *present value of an annuity* can be calculated if we discount each periodic payment separately to the starting point and then add up all the discounted parts.

$$\text{Present Value of an Annuity: } \text{PV}_{\text{annuity}} = \frac{\text{PMT}}{i} * \left[1 - \frac{1}{(1+i)^n}\right] * (1+iT) \tag{30}$$

where

PMT fixed periodic payment, alternatively referred to as cash flow

i interest rate per compounding interval *n*

n = m * t total number of compounding intervals

t time or number of periods

The *future value of an annuity* is equal to the sum of face value of periodic annuity payments and the total compound interest earned on all periodic payments till the future value point. It is calculated by the following formula:

$$\text{Future Value of an Annuity: } \text{FV}_{\text{annuity}} = \frac{\text{PMT}}{i} * [(1+i)^n - 1](1+iT) \tag{31}$$

where

i interest rate per compounding interval *n*

n = m * t total number of compounding intervals

PMT fixed periodic payment
t time or number of periods

The *present value of a growing annuity with constant rate* g ($g \neq i$) is used to calculate the present value of a series of cash flows, or payments that grow at a constant growth rate g. It is calculated by

$$\mathrm{PV_{GA}} = \frac{\mathrm{PMT}}{(i-g)}\left[1 - \frac{(1+g)^n}{(1+g)^n}\right](1+iT) \tag{32}$$

where
PMT fixed periodic payment
i rate per period
g growth rate
n number of compounding intervals
t time unit

The *future value of a growing annuity* with constant rate g ($g \neq i$) is calculated by

$$\mathrm{FV_{GA}} = \frac{\mathrm{PMT}}{(i-g)}[(1+i)^n - (1+g)^n](1+iT) \tag{33}$$

where
PMT fixed periodic payment
i rate per period
g growth rate
n number of compounding intervals
t time unit

A ***perpetuity*** refers to an infinite and constant stream of identical cash flows. The *present value of perpetuity* is the sum of the discounted value of each periodic payment of the perpetuity. *Present value of perpetuity* is finite because the discounted value of far future payments of the perpetuity reduces considerably and reaches close to zero. Therefore, it is calculated by the following formula:

$$\text{Present Value of a Perpetuity: } \mathrm{PV_{perpetuity}} = \frac{A}{R} \tag{34}$$

where
A fixed periodic payment
R interest rate or discount rate per compounding period

A *growing perpetuity* is a series of periodic payments that grow at a constant (proportionate) rate g and are received for an infinite amount of time. The *present value of a growing perpetuity* is the cash flow after the first period divided by the difference between the discount rate and the growth rate:

$$CF_{t+1} = (1+g) * CF_t \tag{35}$$

$$PV_{GP_t} = \frac{D_1}{(R-g)} \tag{36}$$

where
CF cash flow,
$\mathbf{D_1}$ dividend or coupon at period 1,
R discount rate,
g growth rate.

The formula (36) can be used to calculate the *present value of a growing perpetuity* (GP) in commercial real estate. The rental cash flows can be considered as indefinite and will grow over time. In addition, it is noteworthy to mention that the discount rate **R** must be higher than the growth rate **g** when using this formula.

In theory, if the growth rate is higher than the discount rate, the growing perpetuity would have an infinite value.

Inflation has an impact on our consumption and not on our amount of money. Alternatively referring to, it has an impact on our purchasing power and our decision making.

Real return (**RR**) is calculated based on the following formula:

$$\text{Real Return:}\ 1 + \text{RR} = \frac{1+R}{1+\pi} \approx R - \pi \tag{37}$$

where
R nominal return
$\boldsymbol{\pi}$ rate of inflation

Annual percentage rate (**APR**) measures the amount of simple interest earned in a year. It is not considered as a discount rate but is used in order to get a discount rate (such as the *effective annual rate* (**EAR**)).

Annual percentage yield is equal to *effective annual rate*. The latter measures the actual amount of interest earned or paid in a year. It is considered as a discount rate and is what matters for computing interest and discounting cash flows.

The relation between **APR** and **EAR** is represented by the following formula:

$$\text{EAR} = \left(1 + \frac{\text{APR}}{m}\right)^m - 1 = (1 + \text{PIR})^m - 1 \tag{38}$$

where

m number of compounding periods per year

PIR *Periodic interest rate* (PIR) or *periodic discount rate* (PDR)

$$\mathrm{PIR} = \frac{\mathrm{APR}}{m} \tag{39}$$

3.1 Cost of Capital

The *cost of capital* is fundamentally important in both the investment decision making from entity's management as well as the entity's investment valuation by investors. More precisely, if an entity decides to invest on projects with long-term benefits (potential projects generate return in excess of the cost of capital), then the entity has created value; in the opposite direction, the entity has destroyed value.

The *cost of raising capital for a* corporation is usually calculated based on its *cost of debt*, its *cost of equity,* and some other ways of *investment financing, such as preferred stock and convertible bond, if applicable.*

The *cost of capital* is defined as the required rate of return that bondholders and owners require as compensation for their capital investment in the company as well as for the average-risk investment of a company. The most common measure to use is the *weighted average cost of capital* (WACC) [8].

The weights within the WACC calculation are the proportions of the various sources of capital that the entity uses to support its investment operations. WACC is calculated as the weighted average of the after-tax marginal costs of each source of capital, as in the following equation:

$$\mathrm{WACC} = w_\mathrm{d} * r_\mathrm{d}(1-t) + w_\mathrm{p} * r_\mathrm{p} + w_\mathrm{e} * r_\mathrm{e} \tag{40}$$

where

$\mathbf{w_d}$ proportion of debt (used by entity to raise new capital)

$\mathbf{r_d}$ before-tax marginal cost of debt

t entity's marginal tax rate

$\mathbf{w_p}$ proportion of preferred stock

$\mathbf{r_p}$ marginal cost of preferred stock

$\mathbf{w_e}$ proportion of equity

$\mathbf{r_e}$ marginal cost of equity

WACC is extremely useful and is widely used by financial analysts in business valuation. The different components of WACC are calculated via different methods.

The *before-tax cost of debt* is estimated based on the yield to maturity or the bond rating. The *yield to maturity method* is based on the bond valuation equation. It is calculated based on:

$$P_0 = \frac{\mathrm{CF}}{(1+y)} + \frac{\mathrm{CF}}{(1+y)^2} + \cdots + \frac{\mathrm{CF}+A}{(1+y)^n} = \left(\sum_{t=1}^{n} \frac{\mathrm{CF}}{(1+y)^t}\right) + \frac{\mathrm{FV}}{(1+y)^n} \tag{41}$$

where

CF coupon of a bond (fixed payment amount for a prespecified period of time)
A principal or face value
y Yield or Mean Annual Return

It is also important to mention that interest payments are generally tax-deductible, and the after-tax cost of debt is the true, effective cost of debt to the entity.

The *cost of preferred stock* is the preferred stock dividend divided by the current preferred stock price:

$$r_p = \frac{D_p}{P_p} \tag{42}$$

where

$\mathbf{D_p}$ preferred stock dividend
$\mathbf{P_p}$ current preferred stock price

The cost of equity is the rate of return required by the entity's stockholders/equity holders. This cost is mainly estimated via the *CAPM* (usually used by large publicly traded corporations) and in some other cases via the *dividend discount model.* Other methods that are used are the following: *arithmetic average, historical return, multibeta CAPM, investor expectations,* and *regulator decisions*.

The *capital asset pricing model* formula that is used to estimate the *cost of equity* is the following:

$$E(r_i) = r_f + \beta_i[E(r_m) - r_f] \tag{43}$$

where

$\mathbf{E(r_i)}$ expected cost or equity (common stock)
$\mathbf{r_f}$ risk-free rate
$\mathbf{r_m}$ Market/equity risk premium
$\boldsymbol{\beta_i}$ beta

In cases where country and exchange rate risks are not diversified, we can adjust our measure of systematic risk by a *country equity premium* to reflect this no diversifiable risk. In order to calculate the *country equity premium,* we use the following formula:

$$\text{Country Equity Premium} = \text{Country Default Spread} * \frac{\sigma_{\text{equity}}}{\sigma_{\text{sovereign bond}}} \tag{44}$$

where

$\boldsymbol{\sigma_{equity}}$	annualized standard deviation of equity index
$\boldsymbol{\sigma_{sovereign\ bond}}$	annualized standard deviation of the sovereign bond market in terms of the developed market currency

The *dividend discount model approach* is used alternatively to CAPM as a way to calculate the cost of equity via the following formula:

$$r_e = \frac{D_1}{P_0} + g \tag{45}$$

where

$\mathbf{P_0}$ current market value of the financial market index
$\mathbf{D_1}$ next period's expected dividends on the index
$\mathbf{r_e}$ required rate of return on the market
g expected growth rate of dividends

We can also estimate the *sustainable growth rate* in *the dividend discount model* by

$$g = \left(1 + \frac{D}{\text{EPS}}\right) * \text{ROE} \tag{46}$$

where

g growth rate of dividends
D dividend
EPS Earnings per share
ROE return on equity

Alternatively, we can estimate the *growth rate* by using published forecasts of financial analysts.

Other approaches in estimating the *cost of equity* comprise the *bond yield plus risk premium approach*. This approach is based on financial theory that the cost of capital of riskier cash flows is higher than that of less risky cash flows.

Therefore, we sum the before-tax cost of debt and add a risk premium that compensates for the additional risk of equity as opposed the one of debt. The formula to be estimated is the following:

$$r_e = r_d + \text{Risk premium} \tag{47}$$

where
$\mathbf{r_e}$ return on equity
$\mathbf{r_d}$ return on debt

The *risk premium* within the framework of *bond yield plus risk premium approach* is defined as the difference between the cost of equity and the entity's cost of debt. It is a forward-looking measure that reflects the additional risk associated with the shares of the entity as opposed to the bonds of the entity. We usually estimate this premium using historical spreads between bond yields and stock yields.

Under the developed country markets context, a usual risk premium added is within the range of 3–5 %.

Please take into account that these three different approaches of estimating the cost of equity might yield different results. These disparities are not unusual and reflect the difficulty of this estimation.

4 Capital Budgeting

Capital budgeting is the process of making decisions about long-term projects the entity should accept for investment or should reject. This kind of projects have usually a life of one year or more [24].

The typical steps in a capital budgeting process are the following:

- generating ideas,
- analyzing individual proposals,
- planning the capital budget, and
- monitoring and post-auditing.

The candidate projects for the capital budgeting process are as follows:

- Replacement;
- Expansion;
- new products and services; and
- regulatory, safety, and environmental.

These projects are evaluated based on the various methods.

- One of them is the *net present value*. It is considered as the *present value of all after-tax cash flows* and is calculated by the following formula:

$$\mathrm{NPV} = \sum_{t=1}^{T} \frac{\mathrm{CF}_t}{(1+R)^t} \tag{48}$$

where

R required rate of return,

CF cash flow at time **t**, for **t = 1**, …, **T**

One characteristic example of a project valuation for a company that involves WACC for the calculation of NPV is the following:

$$\text{NPV} = \text{PV of inflows} - \text{PV of outflows} \tag{49}$$

Theoretically, projects with positive NPV increase the value of the company and the value of its stock.

- The *internal rate of return* (IRR) is the discount rate that sums up the present value of all future cash flows to zero. The following formula is solved for the IRR:

$$\sum_{t=0}^{T} \frac{\text{CF}_t}{(1+\text{IRR})^t} = 0 \tag{50}$$

The capital budgeting decision rules concern investing when NPV > 0, IRR > R, or when PI > 1.0.

For mutually exclusive projects with different ranking between NPV and IRR, the project with the highest NPV is the one that prevails.

- The *payback period* is based on cash flows and is the number of years needed to recover the original investment in a project. Accordingly, the *discounted payback period* is the number of years required for the *cumulative discounted cash flows* from a project to equal the original investment.
- The *average accounting rate of return* (AAR) is defined as follows:

$$\text{AAR} = \frac{\text{Average net income}}{\text{Average Book Value}} \tag{51}$$

- Finally, the *profitability index* (PI) is defined as the present value of a project's cash flows divided by the initial investment:

$$\text{PI} = \frac{\text{PV of future cash flows}}{\text{Initial investment}} = 1 + \frac{\text{NPV}}{\text{Initial investment}} \tag{52}$$

There are no capital budgeting decision rules for the *payback period*, *discounted payback period,* and *average accounting rate of return*, since they are not considered quite reliable measures.

4.1 Practical Tips in Capital Budgeting

Firms have to invest in projects that by definition exceed their costs. This results in a positive NPV, which is forecasted. However, investment projects that may appear with a positive NPV may be due to forecasting errors. In order to evaluate the influence of forecasting errors on the estimated NPV of the project, several tools exist and are listed below:

- *Sensitivity Analysis*:
  - It provides an analysis of the effect on the estimated NPV when an underlying assumption changes such as market size, market share, or the opportunity cost of capital.
  - It also uncovers the NPV sensitivity to changes in key variables.
- *Scenario Analysis* analyzes the impact on NPV under a particular combination of assumptions. This type of analysis is particularly convenient when the variables are interrelated. A characteristic example can be an economy that enters to a recession due to high oil prices. Indeed, the firm's cost of capital structure, the demand for the product, and the inflation may differ. Therefore, instead of analyzing the effect on NPV of a single variable (as in sensitivity analysis previously), scenario analysis considers the effect on NPV of a consistent combination of variables.
- *Break-even Analysis* analyzes the level at which the company breaks even. This is the point at which the present value of revenues is exactly equal to the present value of total costs. Therefore, break-even analysis tries to respond to the question on how much should the firm sell before the production becomes profitable.
- *Simulation Analysis*
  - The investment project is modeled via Monte Carlo simulation which considers all possible combinations of outcomes. This simulation analysis involves the following three steps:
    - I. We first model the investment project by specifying the project's cash flows as a function of revenues, costs, depreciation, as a function of market size, market shares, unit shares, and costs.
    - II. Later, we specify the probabilities for each of the underlying variables. Some examples include specifying the range for the expected market share as well as other variables in the model.
    - III. Finally, we simulate the cash flows using the model and probabilities assumed above and calculate the NPV.

4.2 *Business Risks*

Business analysts refer to the use of fixed costs in a corporation's capital structure as *leverage* because it influences corporation's earnings. Leverage can magnify earnings both on the upside and the downside.

Business risk is the risk associated with *operating earnings* and refers to both *sales risk* (uncertainty taking into account the price and quantity of sales) and *operational risk* (related to the use of fixed costs in operations). *Financial risk* is the risk associated with the way the corporation finances its operations (choice between debt and equity). There are several *capital structure theories* in corporate finance that attempt to answer to this dilemma of choice.

In particular, the *degree of operating leverage* is the ratio of the percentage change in operating income to the percentage change in units sold. The *degree of financial leverage* is the percentage change in net income for a given percentage change in operating income. Finally, the *degree of total leverage* is a measure of sensitivity of the CFs to owners to changes in unit sales. It is considered as the *degree of operating leverage* plus the *degree of financial leverage* [21].

5 Fundamentals on Payout Policy

Payout policy of the firm constitutes the entity's cash dividend payments and share repurchase policies together. Both contribute to the distribution of the corporation's cash to its shareholders.

Payout policy is considered as second most important for approval, in terms of decision making by the board of directors. Capital expenditures and its way of being finance follows.

The heart of the dividend policy debate concerns the following question: *Should the firm pay out money to its shareholders, or should the firm take that money and invest it for its shareholders?* Therefore, dividend policy can be accounted as a time pattern of dividend payout [13]. More precisely, should the corporation pay out a large percentage of its earnings now, or a small (or even nonexistent) percentage? [16]

Dividends can take different forms: *regular or irregular cash payments*, *stock dividends,* or *stock splits*. The term dividend refers to cash paid out for earnings. Only *cash dividends* are considered as payments for the shareholders. In general, any direct payment by the corporation to the shareholders may be accounted as a dividend or a part of dividend policy.

The main types of cash dividends are the following:

- *regular cash dividends,*
- *extra dividends*: may or may not be repeated in the future,

- *special dividends*: their payment happens once and is not yet repeated,
- *liquidating dividends*: some or all of the business has been liquidated or sold off, and
- *stock repurchase*: another method to pay out a firm's earnings to its owners. In some jurisdictions, it provides favorable tax treatment compared to dividends.

Distribution happens only if a payment is made from sources other than current or accumulated retained earnings [2].

Stock dividends and splits do not create wealth for shareholders. They only serve to split equity into smaller pieces. *Reverse stock splits* usually happen when a stock has reached one of its lowest prices and do not directly affect shareholder wealth. *Regular cash dividends*, as opposed to *irregular cash dividends*, *stock splits* and *stock dividends*, demonstrate a commitment to pay cash to stockholders/equity holders on a regular basis (quarterly, semiannual, or annual).

A *share repurchase* is taken as equivalent to the payment of a cash dividend of equal amount in its effect on shareholder's wealth. The announcement of *share repurchases* is quite often accompanied by positive excess returns in the market when the market price is perceived as reflecting management's view that the stock is undervalued, and earnings per share can increase as a result of fewer shares outstanding [25].

A further initiation of regular cash dividends will also have a positive impact on share value. The management is perceived as having confidence in the future to commit on paying out cash to shareholders. This is also named as the *signaling hypothesis*. In addition, from an investor's side, some institutional, as well as individual, shareholders consider regular cash dividend payments as a measure of investment quality.

Some of the factors that favor a low-dividend payout may be the taxation of dividend income and capital gains and flotation costs. Of course, taxation issues differ among countries. In addition, a corporation may be prohibited by law from paying dividends if the dividend amount exceeds the firm's retained earnings.

On the contrary, there are also other real-world factors that favor a high-dividend payout. Some of them are the following:

- *desire for current income;*
- *resolution of uncertainty*: Gordon stated that a high-dividend policy benefits stockholders because it resolves uncertainty [5]; and
- *tax and legal benefits from high dividends*: Some characteristic examples account for pension funds and university endowment funds. The former have a fiduciary responsibility to invest the money prudently, while the latter are frequently prohibited from spending any of their principal. Therefore, they have strong desire to hold high-dividend yield stocks in order to be able to spend. The previous categorization also includes tax-free institutions as well.

6 The Discounted Cash Flow Approach to Investment Valuation

6.1 Equity Valuation

Equity valuation refers to the valuation of an equity stake of an entity. Some principal categories of models in equity valuation do exist:

- *Present Value Models* or alternatively *discounted cash flow models*, estimate the intrinsic value of a security as the present value of future benefits (dividends) expected to be received from the security. In these types of models, benefits are defined in terms of cash expected to be distributed to shareholders (*DDM* discussed earlier) or in terms of CFs available to be distributed to shareholders after meeting the capital expenditure and working capital needs. The *free-CF-to-equity model* is given by the following formula:

$$\text{Future benefits} = \text{free cash flow}$$

$$\text{Value}_0 = \sum_{t=1}^{\infty} \frac{\text{FCFE}_t}{(1+R)^t} \tag{53}$$

- *Multiplier Models* or alternatively *market multiple models* are based on *share price multiples* or *enterprise value multiples* [22]. *Share price multiples* compare relative values and estimate the intrinsic value of a common share from a price multiple for some fundamental variable, such as *revenues*, *earnings*, *cash flows*, or *book value*. Some of the multiples are the following:

$$\text{price to earnings} = \frac{P}{E} \tag{54}$$

$$\frac{P}{E} = \frac{\text{PV}_0}{\text{EPS}} = \frac{1}{R} + \frac{\text{NPVGO}}{\text{EPS}} \tag{55}$$

where

$\mathbf{PV_0}$ Present value at time 0
R discount factor
NPVGO Net present value of growth opportunities

$$\text{price to sales} = \frac{P}{S} = \frac{\text{share Price}}{\text{Sales per share}} \tag{56}$$

- *comparative approach*

  This approach is characterized as a relative valuation approach. The basic aspect in such an approach is that it should bear some similarity with other equities at the same class and industry. If we take into account a stock, this has to be compared with its key competitors. Differences in value between firms are translated as an opportunity. It actually results that the equity being valued is undervalued and can be bought and held until value increases. The opposite could hold true, which could present opportunity for selling the stock.

  There are two primary comparative approaches:

  I. The first approach checks the market comparables for a firm and for its other competitors. Usual market multiples include *price to earnings* (P/E), *price to book value* (P/B), *price to free cash flow* (P/FCF), *enterprise value to sales* (EV/S), and *enterprise multiple* [9].

  II. The second approach accounts for market transactions where similar type of firms have been bought out, acquired by other competitors, private equity firms, or other institutional investors.

- *Asset-based valuation models* estimate the intrinsic value of a common share from the estimated value of the assets on an entity minus the estimated value of its liabilities and preferred shares. The equation resembles with the principal equation of a balance sheet. The estimated market value of the asset is determined by making adjustments to the book value of assets and liabilities. The academic theory that backs up the asset-based valuation approach considers that the value of the business is equal to the sum of the value of its business's assets.

A financial analyst may compare estimates of *value* and *market price* in order to arrive to the conclusion on whether the asset is fairly valued, overvalued, or undervalued. More precisely,

- an overvalued asset is the one that has an intrinsic value higher than the market value,
- a fairly priced asset is the one that has an intrinsic value equal to the market value, and
- an undervalued asset is the one that has an intrinsic value lower than the market value.

The *value of equity* is taken by discounting the expected cash flows to equity, which means the *residual cash flows* after taking into account all the *expenses, tax obligations, interest, and principal payments*, at the *cost of equity*. The *market value of equity* is given by the following formula:

$$\text{Market Value of Equity} = \sum_{t=1}^{T} \frac{\text{CF to Equity}_t}{(1+r_e)^t} \tag{57}$$

where
CF to Equity$_t$ expected cash flow to equity in period t
r_e cost of equity

The computation of *growth rate of dividends* is also important for ***equity valuation of the firm***. Assume that the company does not grow unless a *net investment* is made. Then, the company needs to retain part of its *earnings* to grow. The calculation of these *earnings* is given by the following formula:

$$\text{Earnings}_{t+1} = \text{Earnings}_t + \text{Retained Earnings}_t * R \tag{58}$$

where
R return on the retained earnings, usually estimated by ROE

In this case, the *sustainable growth rate* is calculated if we divide by earnings as follows:

$$1 + g = 1 + \text{Retention Ratio} * \text{ROE} \tag{59}$$

where

$$\text{Retention Ratio} = \frac{\text{Retained Earnings}}{\text{Earnings}} \tag{60}$$

6.2 *Business Valuation*

Financial analysts usually use the term Business Valuation when they want to value not only the equity of a business but also other claimholders in the firm. Some of the methods they use are the following:

- The *Free Cash Flow for the Firm* (FCFF) is considered as a financial performance measure that shows the net amount of cash that is generated for the firm, consisting of *expenses, taxes, changes in net working capital* (NWC), *and investments* [18]. It is calculated as follows:

$$\begin{aligned} \text{FCFF} = {} & \text{Operating Cash Flow} - \text{Expenses} \\ & - \text{Taxes} - \text{Changes in NWC} - \text{Changes in Investments} \end{aligned} \tag{61}$$

- The *Discounted Cash Flow for the firm* (DCFF) is considered as an essential valuation method that intends to estimate the attractiveness of an investment opportunity (from an investor's point of view). From a firm's point of view, it aims to estimate the intrinsic value of a company [20].

- Based on A. Damodaran's steps, we first estimate the discount rate or rates to use in the valuation. Our discount rate can be either the cost of equity (if we use it to make equity valuation) or cost of capital (if it is used for business valuation). It can also be in real or nominal terms depending on whether the cash flows are in nominal or real terms [17].
- Second, we estimate the current earnings and cash flows on the asset being valued by estimating an expected growth rate in earnings.
- Third, we estimate when the firm will achieve a stable growth rate and what characteristics it will have, in terms of risk and cash flows.
- Finally as a last step, we choose the right DCF model for this asset or corporation and value it.

In order to account for the value of the firm, we discount the expected cash flows of the corporation, which means the residual cash flows after all operating expenses and taxes, prior to debt payments at the WACC. The value of the firm is calculated by the following formula:

$$\text{Value of Entity} = \sum_{t=1}^{T} \frac{\text{CF to Entity}_t}{(1+\text{WACC})^t} \tag{62}$$

where

CF to entity$_t$ Expected cash flow to entity in period t
WACC Weighted average cost of capital

Therefore, as previously mentioned, we discount the cash flow to entity at the cost of capital in order to get the value of the firm, whereas we discount the cash flow to equity at the cost of equity to get the value of the equity.

7 Multiple Choice Questions

1. In the CAPM formula, the risk-free rate is the compensation for the expected loss of the purchasing power.

   a. True
   b. False

2. The debt-to-equity ratio shows the profitability of the firm.

   a. True
   b. False

3. Which are the tools to evaluate the forecasting errors of the estimated NPV? Select from the following (more than one apply):

   a. Discounted payback period
   b. Simulation analysis

c. Payout policy
d. Scenario analysis
e. Business valuation
f. Break-even analysis

4. Leverage is referring to

a. Company's debt
b. Company's equity
c. Both

5. Corporate finance is a field in finance that is dealing with capital budgeting, capital structure, and working capital management functions within a firm.

a. True
b. False

6. John decides to put 100 € in a bank account with interest rate 10 % per year. Which is the future value after 4 years:

a. 234.15 €
b. 139.23 €
c. 146.41 €
d. 187.35 €

7. The company ABC wants to raise €3 million in order to expand its business to other countries. Its corporate tax rate is 20 %. The yield to maturity of its bonds is 6 %.
Calculate the cost of debt for that company.

a. 2.1 %
b. 4.8 %
c. 5.3 %

8. Which of the following consists a potential solution to resolve a shareholders–managers conflict in favor of shareholders?

a. give incentives to the current manager to expand the business activities
b. increase managerial compensation
c. do not do anything

9. Cash dividends are considered as claims for the shareholders.

a. True
b. False

10. The cost of capital accounts for the interactions between corporations and investors in the market.

a. True
b. False

Multiple Choice Answers

1. a,
2. b,
3. b., d., f.,
4. a,
5. a,
6. c,
7. b,
8. c,
9. b,
10. a.

References

1. Barclay MJ, Smith CW (1996) On financial architecture: leverage, maturity and priority. J Appl Corp Finan 8(4):4–17. doi:10.1111/j.1745-6622.1996.tb00679.x
2. Barclay MJ, Smith CW, Watts RL (1995) The determinants of corporate leverage and dividend policies. J Appl Corp Finan 7(4):4–19. doi:10.1111/j.1745-6622.1996.tb00679.x
3. Berk J, DeMarzo P (2013) Corporate finance, 3rd edn. Pearson Education, New York
4. Bhide A (1989) The causes and consequences of hostile takeovers. J Appl Corp Finan 2: 36–59. doi:10.1111/j.1745-6622.1989.tb00178.x
5. Black F, Scholes M (1974) The effects of dividend yield and dividend policy on common stock prices and returns. J Finan Econ 1:1–22. doi:10.1016/0340-405X(74)90006-3
6. Booth L, Cleary WS (2011) Introduction to corporate finance. Wiley, New York (revised)
7. Brealey R, Myers S (2013) Principles of corporate finance. McGraw-Hill, New York
8. Courtois Y (2011) Cost of capital. In: Corporate Finance and Portfolio Management CFA program curriculum, vol 4, Pearson, New York
9. Damodaran A (1994) Investment valuation. Wiley, New York
10. Estrada J (2011) The FT guide to understanding finance: a no-nonsense companion to financial tools and techniques. FT Prentice Hall, Upper Saddle River
11. Estrada J (2010) The essential financial toolkit: everything you always wanted to know about finance but were afraid to ask. Palgrave Macmillan
12. Fabozzi FJ, Wilson RS (1995) Corporate bonds: structures and analysis. Frank J. Fabozzi Associates
13. Fama EE, Babiak H (1968) Dividend policy: an empirical analysis. J Am Stat Assoc 63 (324):1132–1161. doi:10.2307/2285874
14. Goswami G, Noe T, Rebello M (1995) Debt financing under asymmetric information. J Finan 50(2):633–659. doi:10.1111/j.1540-6261.1995.tb04798.x
15. Hamada RS (1972) The effect of firm's capital structure on the systematic risk of common stocks. J Finan 27:435–452. doi:10.2307/2978486
16. Healy P, Palepu K (1988) Earnings information conveyed by dividend initiations and omissions. J Finan Econ 21(2):149–175. doi:10.1016/0304-405X(88)90059-1
17. Howe KM (1992) Capital budgeting discount rates under inflation: a caveat. Finan Pract Educ 2:31–36
18. Jensen MC (1986) Agency costs of free cash flow, corporate finance and takeovers. Am Econ Rev 76:323–329

19. Jensen MC, Meckling WH (1976) Theory of the firm: managerial behavior, agency costs and ownership structure. J Finan Econ v3(4):305–360. doi:10.1016/0304-405X(76)90026-X
20. Keck T, Levengood E, Longfield A (1998) Using discounted cash flow analysis in an international setting: a survey of issues in modeling the cost of capital. J Appl Corp Finan 11 (3):82–99. doi:10.1111/j.1745-6622.1998.tb00505.x
21. Peterson P (2011) Measures of leverage. In: Corporate Finance and Portfolio Management CFA program curriculum, vol 4. Pearson, New York
22. Quiry P, Dallocchio M, Le Fur Y, Salvi A (2009) Corporate finance: theory and practice. Wiley, Chichester
23. Ross S, Westerfield R, Jaff J (2008) Fundamentals of corporate finance, 8th edn (Standard Edition). McGraw-Hill, Irwin
24. Stowe J, Gagné J (2011) Capital budgeting. In: Corporate Finance and Portfolio Management CFA program curriculum, vol 4. Pearson, New York
25. Troughton G (2011) Dividends and shares repurchases: basics. In: Corporate Finance and Portfolio Management CFA program curriculum, vol 4. Pearson, New York
26. Vermaelen T (2005) Share repurchases. Now Publishers, Hanover

Index

C. Machado and J.P. Davim (eds.), *MBA*,
Management and Industrial Engineering, DOI 10.1007/978-3-319-28281-7

D

E

MIX
Papier aus verantwortungsvollen Quellen
Paper from responsible sources
FSC® C105338

If you have any concerns about our products,
you can contact us on
ProductSafety@springernature.com

In case Publisher is established outside the EU,
the EU authorized representative is:
Springer Nature Customer Service Center GmbH
Europaplatz 3, 69115 Heidelberg, Germany

Printed by Libri Plureos GmbH
in Hamburg, Germany